New Heinemann Maths

Answer Book

Contents

Heinemann Educational Publishers
Halley Court, Jordan Hill, Oxford, OX2 8EJ
a division of Reed Educational & Professional Publishing Ltd
www.heinemann.co.uk

Heinemann is a registered trademark of Reed Educational & Professional Publishing Ltd

ISBN 0 435 17881 4

First published 2002

06 05 04 03 02
10 9 8 7 6 5 4 3 2 1

Typeset by Artistix

Printed and bound in Great Britain

1

1 Peter 3 hundreds   Samia 300 thousands
   Gill 30 thousands   Dev 3 million
   Ragia 3 tens   Joan 3 units

2 (a) 2 thousands   (b) 3 tens   (c) 600 thousands
   (d) 4 million   (e) 2 units   (f) 30 thousand
   (g) 7 thousands   (h) 5 hundred   (i) 10 million

3 (a) 3 210 650
   (b) 4 890 395
   (c) 1 980 374
   (d) 3 809 769
   (e) 1 809 536
   (f) 3 980 172

2

1 Game A: Rob   Game B: Computer
   Game C: Computer   Game D: Computer

2 (a) Six million, fifty-four thousand three hundred
   (b) Five million, four hundred thousand and thirty

3 (a) 630 105, 623 150, 562 310, 560 320, 506 123
   (b) 745 585, 745 850, 2 458 407, 2 754 805, 7 850 450

4 (a) 3 500 000   (b) 1 250 000
   (c) 6 950 000   (d) 2 650 000

5 (a) 400 624
   (b) 10 031 089
   (c) 7 605 012

6 (a) Fifty-three thousand
   (b) One hundred and seventy-five thousand and two
   (c) One hundred and twenty thousand, seven hundred and thirteen
   (d) Four million, six hundred thousand, one hundred and fifty
   (e) Six million, eighty thousand, four hundred and ten
   (f) Ten million, three hundred thousand, six hundred and nine

3

1 Marvin 73 625
   Rosie 508 300
   Wayne 1 500 650
   Kirsty 904 009

2 (a) 3 169 035   (b) 3 168 135   (c) 3 268 035
   (d) 4 168 035   (e) 3 168 045   (f) 3 178 035

3 (a) 1 275 453   (b) 1 274 463   (c) 1 275 363
   (d) 1 265 463   (e) 275 463   (f) 1 175 463

4 (a) 327 000   (b) 553 000
   (c) 6 796 500   (d) 930 200

5 (a) 328 000   (b) 378 000
   (c) 219 000   (d) 119 000
   (e) 5 870 000   (f) 3 870 000

4

1 (a) 230   (b) 2300   (c) 670   (d) 67 000
   (e) 15 400   (f) 2090   (g) 681 000   (h) 3560
   (i) 1 300 000 (j) 714 000   (k) 20 650   (l) 1 001 000

2 (a) 180   (b) 6200   (c) 7000
   (d) 39 000   (e) 15 200   (f) 5 260 000
   (g) 100   (h) 1000   (i) 734

3 (a) £37   (b) £104   (c) £43

4 (a) 62   (b) 12   (c) 15
   (d) 520   (e) 70   (f) 590
   (g) 10   (h) 100   (i) 392 000

5 (a) 630 000   (b) 51 000   (c) 9 800 000

5

Pupil's own work.

6

1 Frogger 3000 copies   Invaders 9000 copies
   Breakout 6000 copies   Tetris 5000 copies

2 (a) 7000   (b) 12 000   (c) 30 000   (d) 50 000
   (e) 20 000   (f) 62 000   (g) 132 000   (h) 554 000

3 (a) 300   (b) 500   (c) 2700   (d) 9100
   (e) 34 800   (f) 21 100   (g) 143 200   (h) 320 900

4 Frogger £4 000 000   Invaders £3 000 000
   Breakout £6 000 000   Tetris £2 000 000

5 (a) 3 000 000   (b) 9 000 000   (c) 7 000 000
   (d) 6 000 000   (e) 10 000 000 (f) 1 000 000
   (g) 12 000 000 (h) 15 000 000 (i) 20 000 000

7

1 (a) $60 + 40 + 90 + 80 = \mathbf{270}$
   (b) $70 + 90 + 50 + 70 = \mathbf{280}$
   (c) $60 + 30 + 80 + 50 = \mathbf{220}$
   (d) $80 + 10 + 50 + 40 = \mathbf{180}$

2 (a) 190   (b) 220   (c) 250   (d) 290

3 (a) 90   (b) 95   (c) 146   (d) 173
   (e) 180   (f) 196   (g) 173   (h) 181
   (i) 242   (j) 90   (k) 68

4 $70 + 75 + 77 + 71 = \mathbf{293}$ questions

5 (a) 246   (b) 218   (c) 158   (d) 489

8

1 (a) $155\,km + 319\,km = \mathbf{474\,km}$ (b) $319\,km + 218\,km = \mathbf{537\,km}$
   (c) $218\,km + 246\,km = \mathbf{464\,km}$ (d) $246\,km + 137\,km = \mathbf{383\,km}$

2 (a) $319\,km + 109\,km = \mathbf{428\,km}$ (b) $246\,km + 109\,km = \mathbf{355\,km}$

3 (a) 990   (b) 780   (c) 881
   (d) 882   (e) 582   (f) 695
   (g) 976   (h) 682   (i) 970
   (j) 891   (k) 494   (l) 883

4 (a) $149 + \mathbf{533} = 682$ (b) $519 + \mathbf{237} = 756$
   (c) $\mathbf{275} + 216 = 491$

9

1 (a) 190 + 243 = **433**　　(b) 243 + 480 = **723**
　(c) 480 + 374 = **854**　　(d) 374 + 350 = **724**
　(e) 350 + 568 = **918**

2 (a) 850　　(b) 934　　(c) 716　　(d) 926
　(e) 839　　(f) 867　　(g) 765　　(h) 616
　(i) 549　　(j) 547　　(k) 878　　(l) 909

3 (a) 193 + **560** = 753　　(b) 770 + **175** = 945
　(c) **456** + 453 = 909

4 Number of shoppers between 2.00pm and 3.00pm is 380,
　Number of shoppers between 3.00pm and 4.00pm is 240
　Number of shoppers between 2.00pm and 4.00pm is 380 + 240 = **620**

10

1 (a) £840 + £704 = **£1544**　　(b) £222 + £840 = **£1062**
　(c) £955 + £513 = **£1468**　　(d) £704 + £955 = **£1659**
　(e) £931 + £840 = **£1771**　　(f) £513 + £704 = **£1217**
　(g) £704 + £931 = **£1635**　　(h) £955 + £222 = **£1177**
　(i) £513 + £931 = **£1444**　　(j) £840 + £513 = **£1353**

2 (a) 1190　　(b) 1088　　(c) 1477
　(d) 1078　　(e) 1488　　(f) 1178
　(g) 1259　　(h) 1258　　(i) 650 + **520** = 1170
　(j) **910** + 380 = 1290　　(k) 827 + **821** = 1648

3 (a) £931 + £222 − £29 − £15 = **£1109**
　(b) £840 + £513 − £29 − £15 = **£1309**

11

1 (a) music4u.co　　4986　　(b) music4u.co　　7989
　　sportsnews.co　5938　　　　sportsnews.co　6899
　　gameszone.co　9894　　　　gameszone.co　5599
　　fashion.co　　5997　　　　fashion.co　　8978
　　starsigns.co　6968　　　　starsigns.co　8879
　　freebie.co　　9895　　　　freebie.co　　8579

2 Hawks　　1608 + 1384 = **2992**
　Sharks　　2355 + 1339 = **3694**
　Arrows　　1457 + 2326 = **3783**

3 (a) 6671　　(b) 6981　　(c) 9882

12

1 (a) £28 974　　(b) £47 882
　(c) £46 911　　(d) £39 078
　(e) £40 166　　(f) £31 033

2 (a) S.S. Storm　　79 585 miles　　(b) S.S. Storm　　85 001 miles
　　S.S. Tempest　61 231 miles　　　　S.S. Tempest　80 024 miles
　　S.S. Spray　　42 900 miles　　　　S.S. Spray　　63 511 miles
　　S.S. Coral　　90 224 miles　　　　S.S. Coral　　94 320 miles

3 (a) 94 953　　(b) 92 351　　(c) 81 702
　(d) 52 131　　(e) 97 441　　(f) 93 610

13

1 Mark 9090　　Zoe 5968　　Leela 3466
　Sean 8853　　Holly 3476　　Brad 6600

2 Team A 3664   Team B 6822   Team C 6003

3 **(a)** 3790   **(b)** 8946   **(c)** 2020
(d) 2596   **(e)** 12865   **(f)** 15390

4 **(a)** $6 + 8135 + \mathbf{95} + 364 = 8600$ **(b)** $72 + \mathbf{106} + 6058 + 4 = 6240$

14

1 **(a)** 1st row; $6 + 5754 + 739 + 13146 + 82 = 19727$
2nd row; $36 + 3 + 6108 + 537 + 12908 = 19592$
3rd row; $12507 + 72 + 8 + 5956 + 459 = 19002$
4th row; $210 + 11365 + 54 + 9 + 7835 = 19473$
5th row; $8879 + 893 + 13077 + 61 + 5 = 22915$
The 5th row has the greatest total.

(b) 1st column; $6 + 36 + 12507 + 210 + 8879 = 21638$
2nd column; $5754 + 3 + 72 + 11365 + 893 = 18087$
3rd column; $739 + 6108 + 8 + 54 + 13077 = 19986$
4th column; $13146 + 537 + 5956 + 9 + 61 = 19709$
5th column; $82 + 12908 + 459 + 7835 + 5 = 21289$
The 2nd column has the smallest total.

2 **(a)** 17368 $(3 + 36 + 210 + 5754 + 11365)$
(b) 23009 $(9 + 82 + 893 + 8879 + 13146)$

3 13146 and 54

4 $\mathbf{537} + \mathbf{459} + \mathbf{3} = 999$

5 Pupil's own answers.

6 $\mathbf{5754} + \mathbf{7835} + \mathbf{8879} = 22468$

15

1 **(a)** 470   **(b)** 180   **(c)** 270   **(d)** 350

2 **(a)** $550 - 260 = \mathbf{290}$ **(b)** $640 - 470 = \mathbf{170}$ **(c)** $420 - 180 = \mathbf{240}$
(d) $930 - 370 = \mathbf{560}$ **(e)** $810 - 230 = \mathbf{580}$ **(f)** $740 - 560 = \mathbf{180}$
(g) $730 - \mathbf{350} = 380$ **(h)** $\mathbf{920} - 540 = 290$ **(i)** $860 - \mathbf{290} = 570$

3 **(a)** $527 - 340 = \mathbf{187}$ **(b)** $613 - 260 = \mathbf{353}$
(c) $752 - 480 = \mathbf{272}$ **(d)** $527 - 480 = \mathbf{47}$
(e) $613 - 340 = \mathbf{273}$ **(f)** $752 - 260 = \mathbf{492}$

4 **(a)** 164   **(b)** 356   **(c)** 185
(d) 285   **(e)** 293   **(f)** 388
(g) 551   **(h)** 462   **(i)** 276
(j) 589   **(k)** 168   **(l)** 151

5 **(a)** $736 - \mathbf{180} = 556$ **(b)** $951 - \mathbf{690} = 261$ **(c)** $611 - \mathbf{330} = 281$
(d) $\mathbf{304} - 120 = 184$ **(e)** $\mathbf{627} - 360 = 267$ **(f)** $\mathbf{854} - 180 = 674$

16

1 **(a)** £265   **(b)** £263
(c) £474   **(d)** £237
(e) £653   **(f)** £187
(g) £315   **(h)** £84

2 **(a)** 378   **(b)** 166   **(c)** 372
(d) 329   **(e)** 388   **(f)** 243
(g) 555   **(h)** 274   **(i)** 437
(j) 186   **(k)** 472   **(l)** 261

3 **(a)** 634 pages   **(b)** 227 pages
(c) 383 pages   **(d)** 282 pages

17

1 (a) £323   (b) £438
   (c) £538   (d) £345
   (e) £355   (f) £615
   (g) £329   (h) £228

2 (a) 346   (b) 319   (c) 617
   (d) 527   (e) 236   (f) 249
   (g) 218   (h) 217   (i) 537
   (j) 513   (k) 226   (l) 127

3 (a) 217   (b) 226   (c) 338

18

1 Firth   3253
   Boulter   2233
   Ainley   7422
   Sorn   6216
   Gilton   6861
   Rebway   5502

2 (a) 3310   (b) 312   (c) 3444
   (d) 1111   (e) 1999   (f) 1805

3 Firth   3418
   Boulter   2149
   Ainley   6318
   Sorn   5239
   Gilton   8129
   Rebway   4115

19

1 (a) £55 400   (b) £23 820
   (c) £42 285   (d) £13 115
   (e) £18 465   (f) £36 935

2 (a) 15 123   (b) 19 825   (c) 25 915
   (d) 23 789   (e) 8885   (f) 36 574

3 £750 465

4 (a) 23 673   (b) 39 148
   (c) 829 704   (d) 305 986
   (e) 50 536   (f) 228 880
   (g) 643 139   (h) 901 806
   (i) 994 038

20

1 190 and 520

2 (a) $8000 - 2700 = 5300 - 1500 = 3800 - 2250 = 1550 - 600 = 950 - 384 = 566$
   (b) $6000 - 3524 = 2476 - 1258 = 1218 - 400 = 818 - 209 = 609 - 440 = 169$

3 (a) 1276   (b) 8032   (c) 1239
   (d) 8041   (e) 6786   (f) 6760

4 211 and 132   256 and 177   290 and 211

5 (a) 19 587   (b) 93 666

6 (a) $6715 - 5306 = 1409$ (b) $93 265 - 47 251 = 46 014$

21

1 (a) £96   (b) £54   (c) £276
  (d) £99   (e) £120   (f) £81
  (g) £368   (h) £189   (i) £165

2 (a) 87   (b) 134   (c) 380   (d) 356
  (e) 228   (f) 322   (g) 520   (h) 162

3 (a) 180   (b) 162
  (c) 544   (d) 304
  (e) 204   (f) 552

4 252

5 £477

22

1 (a) 170   (b) 310   (c) 630   (d) 720
  (e) 490   (f) 270   (g) 350   (h) 400

2 (a) 220   (b) 810   (c) 380   (d) 210

3 (a) 180   (b) 270

4 (a) 420   (b) 540

5 (a) 208   (b) 306   (c) 462   (d) 444
  (e) 378   (f) 528   (g) 496   (h) 756

6 (a) 264   (b) 522

7 (a) 1300   (b) 1900   (c) 2100   (d) 2700
  (e) 600   (f) 900   (g) 1200   (h) 1400

8 (a) 25, 32   (b) 26, 50 or 52, 25

23

1 (a) £1400, £1414, £1386
  (b) £2300, £2323, £2277

2 (a) 3400   (b) 1717   (c) 1584
  (d) 2828   (e) 1900   (f) 3131
  (g) 3366   (h) 2600   (i) 2772

3 (a) Blue £650, Green £663, Red £637
  (b) Blue £1200, Green £1224, Red £1176
  (c) Blue £950, Green £969, Red £931

4 (a) 900   (b) 612   (c) 1274
  (d) 833   (e) 1050   (f) 1377

5 (a) £1862   (b) £1350   (c) £4554
  (d) £1323   (e) £1938   (f) £2929

24

1 (a) £196   (b) £153   (c) £210
  (d) £138 + £68 = £206
  (e) £224 + £105 = £329
  (f) £180 + £288 = £468

2 £348 + £91 = £439

3 (a) 650   (b) 867   (c) 1666

4 £8·10 + £4·48 + £28·42 + £6·75 = £47·75

25

1 (a) £20 832   (b) £21 054
   (c) £21 780   (d) £5034
   (e) £49 248   (f) £44 796
   (g) £22 272   (h) £77 040

2 (a) £10 944 + £13 680 = £24 624
   (b) £28 890 + £33 705 = £62 595
   (c) £28 072 + £28 072 = £56 144
   (d) £17 856 + £13 392 = £31 248

3 India $(2 \times 5472 + 3 \times 2736 = 19\,152)$

26

1 (a) 3276   (b) 7700   (c) 12 508   (d) 9936
   (e) 9367   (f) 6156   (g) 9287   (h) 9702

2 £9906

3 7866 g

4 (a) 13 984 headphones
   (b) 15 813 headphones

27

1 (a) 7   (b) 6   (c) 5   (d) 9   (e) 5   (f) 4

2 (a) 7   (b) 10   (c) 10   (d) 5   (e) 7   (f) 9
   (g) 9   (h) 3   (i) 4   (j) 5   (k) 10   (l) 9

3 (a) $\frac{21}{3} = 7$ (b) $\frac{72}{9} = 8$ (c) $\frac{30}{6} = 5$ (d) $\frac{48}{6} = 8$ (e) $\frac{60}{6} = 10$

   (f) $\frac{32}{8} = 4$ (g) $\frac{24}{8} = 3$ (h) $\frac{16}{4} = 4$ (i) $\frac{36}{6} = 6$ (j) $\frac{90}{10} = 9$

4 (a) 6 each, 2 left over
   (b) 5 each, 3 left over
   (c) 8 each, 3 left over
   (d) 7 each, 5 left over
   (e) 7 each, 4 left over

5 (a) 8 r 8   (b) 6 r 5   (c) 8 r 6   (d) 9 r 1
   (e) 7 r 3   (f) 3 r 2   (g) 10 r 3   (h) 0 r 7
   (i) 10 r 2   (j) 8 r 2   (k) 10 r 6   (l) 2 r 6

6 48

28

1 (a) £30   (b) £90   (c) £90

2 (a) 40   (b) 80   (c) 90   (d) 60
   (e) 90   (f) 90   (g) 70   (h) 70
   (i) $280 \div 7 = 40$ (j) $720 \div 9 = 80$
   (k) $640 \div 8 = 80$ (l) $210 \div 3 = 4070$

3 (a) 13 laps   (b) 12 laps   (c) 15 laps   (d) 13 laps

4 (a) 13   (b) 13   (c) 12   (d) 13   (e) 11   (f) 14
   (g) 12   (h) 19   (i) 12   (j) 17   (k) 15   (l) 13

5 (a) 212   (b) 109   (c) 91   (d) 81   (e) 53
   (f) 91   (g) 104   (h) 81   (i) 104   (j) 101

29

1 (a) 852 stickers   (b) 964 badges   (c) 408 posters
   (d) 954 leaflets   (e) 879 letters

2 (a) 1563   (b) 1135   (c) 4379   (d) 1093
   (e) 1522   (f) 2064   (g) 1296   (h) 1401

3 (a) 837 each, 5 left over   (b) 481 each, 5 left over
   (c) 733 each, 1 left over

4 (a) 893 r 3   (b) 397 r 1   (c) 1634 r 1   (d) 1578 r 1
   (e) 1291 r 4   (f) 1025   (g) 1013 r 2   (h) 1201 r 1

30

1 (a) 23   (b) 25   (c) 32   (d) 31

2 (a) 35   (b) 26   (c) 34   (d) 46

3 (a) 19 in each, 8 left over   (b) 25 in each, 9 left over
   (c) 35 in each, 11 left over

4 (a) 24 r 3   (b) 42 r 4   (c) 45 r 11   (d) 32 r 2
   (e) 43 r 9   (f) 31 r 4   (g) 42 r 9   (h) 40 r 16

31

1 (a) Will (992 ÷ 31)   (b) Mike (936 ÷ 26)
   (c) Devon (924 ÷ 28)   (d) Terry (630 ÷ 18)

2 Pupil's own clue with the answer 34.

3 636 bars

4 (a) 35 Sprinters   (b) 39 Townlinks
   (c) 27 Shuttles   (d) 22 Longhauls

5 24 planes

32

1 (a) 24, 30, 36, 42, 48, 54, 60, 66, 72, 78, 84, 90, 96
   (b) 3, 11, 19, 27, 35, 43, 51, 59, 67
   (c) 72, 65, 58, 51, 44, 37, 30, 23, 16, 9
   (d) 94, 83, 72, 61, 50, 39, 28, 17, 6
   (e) 42, 51, 60, 69, 78, 87, 96
   (f) 257, 232, 207, 182, 157, 132, 107, 82, 57, 32, 7

2 (a) 10, 25, 40, 55, **70, 85, 100, 115** → add 15
   (b) 5, 12, 19, 26, **33, 40, 47, 54** → add 7
   (c) 79, 73, 67, 61, **55, 49, 43, 37** → subtract 6
   (d) 263, 242, 221, 200, **179, 158, 137, 116** → subtract 21
   (e) 25, 44, 63, 82, **101, 120, 139, 158** → add 19
   (f) 92, 84, 76, 68, **60, 52, 44, 36** → subtract 8
   (g) 10, 9·5, 9, 8·5, **8, 7·5, 7, 6·5** → subtract 0·5
   (h) ⁻9, ⁻6, ⁻3, 0, **3, 6, 9, 12** → add 3
   (i) 27, 18, 9, 0, **⁻9, ⁻18, ⁻27, ⁻36** → subtract 9
   (j) 1, 2, 4, 8, **16, 32, 64, 128** → multiply by 2

3 (a) 5, 16, 27, **38, 49**, 60 → add 11
   (b) 79, 64, 49, **34, 19**, 4 → subtract 15
   (c) 13, 38, 63, **88, 113**, 138 → add 25
   (d) 122, **103, 84**, 65, 46, 27 → subtract 19
   (e) 21, 10, ⁻1, ⁻12, ⁻23, ⁻34 → subtract 11
   (f) 36, 15, ⁻6, ⁻27, ⁻48, ⁻69 → subtract 21
   (g) **12**, 31, 50, **69**, 88, 107 → add 19
   (h) 3, 6, 12, 24, **48, 96** → multiply by 2

33

1 (a) 1, 9, 16, 49, 64

(b) 16, 64

(c) 5, 17, 33

2 (a) 1, 4, 9, 16, 25, 36, 49, 64, 81, 100, 121, 144

(b) 169, 196, 225, 256, 289, 324, 361, 400

3 (a) • 1 + 64 = 65   • 25 + 225 = 250

(b) 9 + 16 = 25, 36 + 64 = 100, 81 + 144 = 225, 144 + 256 = 400

4 (a) • 5   • 9

(b) • 31   • 52

5 (a) $2^2 - 1^2 = 4 - 1 = 3 = 2 + 1$

$3^2 - 2^2 = 9 - 4 = 5 = 3 + 2$

$4^2 - 3^2 = 16 - 9 = 7 = 4 + 3$

$5^2 - 4^2 = 25 - 16 = 9 = 5 + 4$

$6^2 - 5^2 = 36 - 25 = 11 = 6 + 5$

(b) • 17   • 19

(c) The difference between the squares of two numbers is equal to the sum of the two numbers.

34

1 (a) 1, 3, 6, 10

(b)

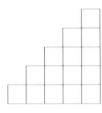

15

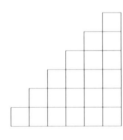

21

(c) 28, 36

(d) Add on nine to the number of boxes in the eighth stack. 36 + 9 = 45

(e) 1, 3, 6, 10, 15, 21, 28, 36, 45, 55, 66, 78, 91, 105, 120

2 (a) Pupil's own work.

(b) Square numbers.

3 (a) 1, 36

(b) • 45 + 55 = 100   • 1 + 21 + 78 = 100

• 3 + 6 + 91 = 100   • 6 + 28 + 66 = 100

35

1 (a) 52, 54, 56, 58

(b) 7, 9, 11, 13

2 (a) 32   (b) 60   (c) 100

(d) 12   (e) 200   (f) 80

3 (a) • even   • even

(b) The product of two even numbers is an even number.

4 (a) 27   (b) 77   (c) 45

(d) 57   (e) 99   (f) 95

5 The product of two odd numbers is an odd number.

6 Children should find that the product of an odd number and an even number is an even number.

36

1 (a) ⁻10°C, ⁻5°C, ⁻2°C, 6°C, 10°C   (b) ⁻8°C, ⁻4°C, 0°C, 3°C, 7°C

(c) 26°C, 1°C, ⁻1°C, ⁻14°C, ⁻27°C   (d) 24°C, 2°C, ⁻2°C, ⁻34°C, ⁻43°C

2 (a) ⁻10°C   (b) ⁻25°C

3 (a) 13°C   (b) 2°C
(c) 0°C   (d) ⁻20°C

4 (a) ⁻14°C   (b) ⁻8°C
(c) ⁻22°C   (d) 11°C

5 (a) 18°C   (b) 36°C   (c) 30°C

6

Mon	Tue	Wed	Thur	Fri	Sat
2·04 m	2·02 m	1·97 m	1·95 m	2 m	1·96 m

7 (a) 5 + 1 = 6   (b) 7 + ⁻1 = 6
    4 + 2 = 6   8 + ⁻2 = 6
    3 + 3 = 6   9 + ⁻3 = 6
                    and so on.

37

1 (a) 168, 504, 1824, 672, 348, 256, 700, 480, 1912, 936
(b) Divide the last two digits by 2. If this number is even then the original number is divisible by 4.

2 (a) 168, 504, 1824, 672, 256, 480, 1912, 936
(b)

Numbers exactly divisible by 8	168	504	1824	672	256	480	1912	936
Half of the number	84	252	912	336	128	240	956	468
Is the half exactly divisible by 4?	Yes	Yes	Yes	Yes	Yes	Yes	Yes	Yes

(c) A number is exactly divisible by 8 if half the number is divisible by 4.

3 (a) 96, 246, 303, 648, 810, 951, 534, 174, 429, 342, 1668
(b) If the sum of the digits is a multiple of 3, the number is exactly divisible by 3.

4 (a) 96, 246, 648, 810, 534, 174, 342, 1668
(b) Rosa's statement is true.
(c) If a number is even and is exactly divisible by 3 then it is divisible by 6.

5 (a) 432, 440
(b) 430, 431, 433, 434, 435, 436, 437, 439, 440

38

1 • 32, 48, 56, 80, 92
• 50, 65, 70, 80, 85, 95

2 (a) • 7, 14, 21, 28, 35, 42, 49, 56, 63, 70
• 9, 18, 27, 36, 45, 54, 63, 72, 81, 90
(b) 21, 42, 63
(c) 36, 72

3 (a) • even numbers   • multiples of 3
(b) 6, 12, 18, 24, 30, 36
(c) Even numbers, multiples of 6

4 (a) • 2, 4, 6, 8, 10, 12, 14, 16, 18, 20, 22, 24, 26, 28, 30, 32, 34, 36, 38, 40
• 5, 10, 15, 20, 25, 30, 35, 40, 45, 50, 55, 60, 65, 70, 75, 80, 85, 90, 95, 100
(b) 10, 20, 30, 40, 50, 60, 70, 80, 90, 100
(c) 10

5 (a) 14   (b) 30   (c) 24   (d) 35
(e) 20   (f) 24   (g) 30   (h) 36

6 (a) green

(b) 47th

7 (a) orange

(b) 33rd

39

1 (a) 15, 20, 30, 40, 45   **(b)** 12, 16, 20, 32, 36, 40

(c) 12, 18, 30, 36   **(d)** 16, 32, 40

(e) 18, 27, 36, 45

2 (a) 1 row of 12, 2 rows of 6, 4 rows of 3, 6 rows of 2, 12 rows of 1

(b) 1 and 12, 2 and 6, 3 and 4

3 (a) 1 and 18, 2 and 9, 3 and 6

(b) 1 and 20, 2 and 10, 4 and 5

(c) 1 and 30, 2 and 15, 3 and 10, 5 and 6

(d) 1 and 36, 2 and 18, 3 and 12, 4 and 9, 6 and 6

4 (a) Square numbers

(b) 9: 1, 3, 9

16: 1, 2, 4, 8, 16

25: 1, 5, 25

64: 1, 2, 4, 8, 16, 32, 64

81: 1, 3, 9, 27, 81

100: 1, 2, 4, 5, 10, 20, 25, 50, 100

(c) Square numbers have an odd numbers of factors.

5 (a) • any two from 6, 8, 10, 14, 15, 21, 22, 26, 27

• any two from 12, 18, 20, 28

• any two from 2, 3, 5, 7, 11, 13, 17, 19, 23, 29

(b) prime numbers

6 (a) 1, 2, 3, 4, 6, 8, 12, 24

(b) 2, 3

(c) $2 \times 2 \times 2 \times 3$

7 (a) $12 = 2 \times 2 \times 3$

(b) $18 = 2 \times 3 \times 3$

(c) $30 = 2 \times 3 \times 5$

(d) $36 = 2 \times 2 \times 3 \times 3$

40

1 (a) T   **(b)** F

(c) T   **(d)** F

(e) T   **(f)** F

(g) T

2 Anna: CA 675 BT

Chris: B 15 TD

Ewan: NE 329 KP

Bianca: G 64 DB

Danny: EH 84 CL

Finlay: AB 144 YP

3 Various answers are possible, including

(a) odd number, square number, factors are 1, 3, 9, 27, 81

(b) odd number, triangular number, a common multiple of 5 and 9

(c) even number, one of its factor pairs is 4 and 54

41

1 (a) $\frac{1}{5} = \frac{2}{10}$ **(b)** $\frac{1}{8} = \frac{3}{24}$

(c) $\frac{4}{5} = \frac{12}{15}$ **(d)** $\frac{3}{7} = \frac{12}{28}$

(e) $\frac{7}{9} = \frac{35}{45}$ **(f)** $\frac{12}{12} = \frac{24}{24}$

2 (a) $\frac{1}{2} = \frac{5}{10}$ **(b)** $\frac{1}{5} = \frac{4}{20}$ **(c)** $\frac{1}{10} = \frac{10}{100}$ **(d)** $\frac{1}{7} = \frac{3}{21}$

(e) $\frac{9}{12} = \frac{3}{4}$ **(f)** $\frac{10}{15} = \frac{2}{3}$ **(g)** $\frac{4}{14} = \frac{2}{7}$ **(h)** $\frac{40}{90} = \frac{4}{9}$

(i) $\frac{3}{6} = \frac{15}{30}$ **(j)** $\frac{42}{56} = \frac{6}{8}$ **(k)** $\frac{5}{10} = \frac{50}{100}$ **(l)** $\frac{60}{80} = \frac{3}{4}$

3 (a) $\frac{2}{6}$ **(b)** $\frac{2}{18}$ **(c)** $\frac{70}{80}$

(d) $\frac{90}{100}$ **(e)** $\frac{24}{36}$ **(f)** $\frac{16}{40}$

4 Pupil writes 3 fractions equal to:

(a) $\frac{1}{4}$ **(b)** $\frac{3}{5}$ **(c)** $\frac{4}{7}$ **(d)** $\frac{3}{3}$

42

1 (a) $\frac{3}{6} = \frac{1}{2}$ **(b)** $\frac{2}{12} = \frac{1}{6}$ **(c)** $\frac{6}{15} = \frac{2}{5}$ **(d)** $\frac{8}{12} = \frac{2}{3}$

(e) $\frac{6}{7} = \frac{30}{35}$ **(f)** $\frac{8}{9} = \frac{24}{27}$ **(g)** $\frac{3}{4} = \frac{15}{20}$ **(h)** $\frac{6}{10} = \frac{60}{100}$

(i) $\frac{24}{48} = \frac{4}{8}$ **(j)** $\frac{1}{4} = \frac{25}{100}$ **(k)** $\frac{40}{50} = \frac{4}{5}$ **(l)** $\frac{6}{9} = \frac{48}{72}$

2 (a) $\frac{2}{3}$ **(b)** $\frac{3}{8}$ **(c)** $\frac{3}{4}$

(d) $\frac{8}{10}$ **(e)** $\frac{3}{9}$ **(f)** $\frac{2}{7}$

3 (a) $\frac{20}{30}, \frac{16}{24}$ **(b)** $\frac{60}{80}$ **(c)** $\frac{80}{100}, \frac{20}{25}, \frac{36}{45}$

4 (a) $\frac{1}{2}$ **(b)** $\frac{1}{3}$ **(c)** $\frac{7}{10}$ **(d)** $\frac{5}{8}$ **(e)** $\frac{5}{9}$

(f) $\frac{1}{3}$ **(g)** $\frac{1}{7}$ **(h)** $\frac{2}{5}$ **(i)** $\frac{5}{6}$ **(j)** $\frac{7}{8}$

5 (a) Bea won $\frac{6}{30}$ or $\frac{1}{5}$ **(b)** Ayub won $\frac{10}{30}$ or $\frac{1}{3}$

(c) $\frac{14}{30}$ or $\frac{7}{15}$ were drawn

6 (a) $\left(\frac{25}{50}\right) \frac{1}{2}$ **(b)** $\left(\frac{10}{50}\right) \frac{1}{5}$ **(c)** $\left(\frac{15}{50}\right) \frac{3}{10}$

43

1 (a) $<$ **(b)** $>$ **(c)** $>$ **(d)** $<$

2 Pupil writes 3 fractions:
(a) greater than one sixth **(b)** smaller than one third

3 (a) half, twice
(b) half, twice
(c) twice, half **(d)** three, one third **(e)** ten, one tenth

4 (a) $1\frac{1}{12}, 1\frac{1}{2}, 1\frac{2}{3}, 2\frac{1}{3}, 2\frac{7}{12}$ **(b)** $3\frac{1}{2}, 3\frac{3}{10}, 2\frac{9}{10}, 2\frac{3}{4}, 2\frac{3}{5}$

5 $\frac{1}{2}$ at 15th mark, $\frac{3}{10}$ at 9th mark, $\frac{2}{3}$ at 20th mark, $\frac{1}{5}$ at 6th mark,

$\frac{9}{10}$ at 27th mark, $\frac{1}{6}$ at 5th mark, $\frac{11}{15}$ at 22nd mark

6 (a) $\frac{7}{10}$ **(b)** $\frac{1}{2}$ **(c)** $3\frac{3}{8}$ **(d)** $4\frac{9}{16}$

7 Pupil finds a fraction that is:

(a) greater than $\frac{1}{2}$ and less than $\frac{3}{5}$

(b) less than $\frac{1}{3}$ and greater than $\frac{1}{4}$

44

1 (a) 15 **(b)** 6 **(c)** 12 **(d)** 5 **(e)** 25 **(f)** 20

2 (a) 27 **(b)** 4 **(c)** 28 **(d)** 6 **(e)** 18 **(f)** 24

3 (a) 6 **(b)** 21 **(c)** 36 **(d)** 14
(e) 8 **(f)** 18 **(g)** 35 **(h)** 16

4 (a) 20   (b) 33   (c) 210   (d) 120
(e) 500   (f) 80   (g) 250   (h) 120

5 (a) 20p   (b) 900 m   (c) 300 g
(d) 17 cm   (e) 90 m   (f) 50 ml

6 (a) 4 m   (b) 63 km   (c) 24 ℓ   (d) 21 kg
(e) 50p   (f) 1 m 60 cm   (g) 5 ℓ 600 ml   (h) 1 kg 750 g

7 $\frac{3}{10}$, $\frac{71}{100}$, $2\frac{9}{10}$

8 (a) $(\frac{70}{100})$ $\frac{7}{10}$ (b) $\frac{9}{20}$ $(\frac{45}{100})$ (c) $1\frac{2}{5}$ $(\frac{140}{100})$

(d) $\frac{1}{5}$ $(\frac{200}{1000})$ (e) $\frac{4}{5}$ $(\frac{800}{1000})$ (f) $\frac{3}{5}$ $(\frac{600}{1000})$ (g) $\frac{1}{25}$ $(\frac{40}{1000})$

45

1 Yellow snail $\frac{627}{1000}$ m

Red snail $\frac{371}{1000}$ m

Blue snail $\frac{789}{1000}$ m

Green snail $\frac{223}{1000}$ m

2 (a) 645 mm   (b) 452 mm   (c) 2 m 934 mm   (d) 3 m 578 mm
(e) 5 m 69 mm

3 (a) $\frac{817}{1000}$ km   (b) $\frac{304}{1000}$ km   (c) $\frac{87}{1000}$ km

4 (a) 739 m   (b) 48 m   (c) 4 km 488 m   (d) 1 km 196 m
(e) 6 km 9 m

5 Hamster $\frac{150}{1000}$ kg,   Gerbil $\frac{108}{1000}$ kg,   Rat $\frac{341}{1000}$ kg,   Mouse $\frac{93}{1000}$ kg

6 (a) 764 g   (b) 22 g   (c) 3 kg 999 g   (d) 7 kg 866 g   (e) 2 kg 7 g

46

1 (a) T   (b) T   (c) F   (d) T   (e) F

2 Pupil's own statements such as:
(a) There is 1 red slab for every 2 grey slabs.
The number of red slabs is half the number of grey slabs.

(b) There is 1 red slab for every 4 grey slabs.
The number of red slabs is $\frac{1}{4}$ of the number of grey slabs.

3 (a) 5 red slabs to 1 grey slab (20 red to 4 grey)

(b) $\frac{1}{5}$

(c) Grey $\frac{1}{6}$,   red $\frac{5}{6}$

4 (a) 6 bags of sand   (b) 15 bags of grit

5 12 buckets of cement, 18 buckets of sand.

47

1 (a) 0, 0·1, 0·2, 0·3, 0·4, 0·5, 0·6, 0·7
(b) 2, 2·5, 3, 3·5, 4, 4·5, 5, 5·5
(c) 4, 4·25, 4·5, 4·75, 5, 5·25, 5·5, 5·75
(d) 10, 9·9, 9·8, 9·7, 9·6, 9·5, 9·4, 9·3
(e) 8, 7·5, 7, 6·5, 6, 5, 5·5, 5, 4·5
(f) 6, 5·75, 5·5, 5·25, 5, 4·75, 4·5, 4·25

2 (a) 0·92, 0·94, 0·96, **0·98**, **1·00**, **1·02**, 1·04
(b) 1·76, 1·77, 1·78, **1·79**, **1·80**, **1·81**, 1·82
(c) 2·54, 2·52, 2·50, **2·48**, **2·46**, **2·44**, 2·42
(d) 3·04, 3·03, 3·02, **3·01**, **3·00**, **2·99**, 2·98

3 (a) 6·47   (b) 7·36   (c) 5·85

(d) 3·25   **(e)** 8·20   **(f)** 1·70

4 **(a)** 0·48   **(b)** 3·59   **(c)** 6·00

5 Any 2-place decimal fraction between
  (a) 0·67 and 0·74   **(b)** 1·81 and 1·9   **(c)** 3·9 and 4·05

6 **(a)** $24\frac{7}{10}$ **(b)** $2\frac{41}{100}$

  (c) $5\frac{89}{100}$ **(d)** $78\frac{3}{10}$

  (e) $50\frac{2}{10}$ **(f)** $3\frac{63}{100}$

  (g) $47\frac{5}{10}$ **(h)** $6\frac{5}{100}$

48

1 **(a)** 3·24   **(b)** 1·75   **(c)** 97·4   **(d)** 45·8
  (e) 13·9   **(f)** 8·07   **(g)** 6·38   **(h)** 2·63

2 **(a)** 1 tenth   **(b)** 6 units   **(c)** 7 hundredths
  (d) 4 tens   **(e)** 2 hundredths   **(f)** 0 tenths

3 **(a)** 2·84   **(b)** 6·76   **(c)** 3·21

4 **(a)** 6·15   **(b)** 8·57   **(c)** 4·3

5 **(a)** 0·40, 0·42, 0·53, 0·58, 0·61
  (b) 0·06, 0·09, 0·1, 0·14, 0·2
  (c) 4·36, 4·63, 5·99, 6·06, 6·34, 6·43
  (d) 0·19, 0·91, 9·01, 9·09, 9·1, 9·9
  (e) 0·8, 0·75, 0·73, 0·7, 0·62
  (f) 1·08, 1·05, 1·0, 0·98, 0·9
  (g) 9·95, 9·75, 9·57, 7·95, 7·79, 7·59
  (h) 8·8, 8·2, 8·08, 8·02, 0·82, 0·28

49

1 **(a)** 5·6 m   **(b)** 4·3 m   **(c)** 5·4 m   **(d)** 5·0 m

2 Gio 3·7 m   Franco 4·2 m   Bert 3·5 m   Dani 4·0 m

3

	Throw 1	Throw 2	Throw 3
Ellie	61 m	57 m	67 m
Cara	73 m	75 m	71 m
Alana	64 m	75 m	68 m

4 **(a)** Harry 16 m + 18 m = 34 m
     Imo 12 m + 20 m = 32 m
  (b) Harry 18 m − 16 m = 2 m
     Imo 20 m − 12 m = 8 m

5 **(a)** Harry 34·71 m   Imo 32·11 m
  (b) Harry 1·89 m   Imo 7·59 m

50

1 **(a)** red 0·97 m   blue 0·76 m   green 0·99 m
     yellow 1·16 m   orange 1·33 m   purple 1·24 m
  (b) red 1·03 m   blue 1·24 m   green 1·01 m
     yellow 0·84 m   orange 0·67 m   purple 0·76 m

2 **(a)** 1·1   **(b)** 0·6   **(c)** 1·7
  (d) 2·9   **(e)** 3·7   **(f)** 1·7

3 **(a)** 5·87   **(b)** 7·89   **(c)** 9·76   **(d)** 8·47   **(e)** 6

4 **(a)** 5·62 + **0·38** = 6   **(b)** **0·82** + 7·18 = 8   **(c)** 6·52 + **0·08** = 6·6
  (d) 2·14 + **0·06** = 2·2   **(e)** **3·43** + 1·32 = 4·75   **(f)** 2·47 + **2·4** = 4·87

5 Pupil's own stories using
   (a) any two of 1·06, 2·7, 0·4, 6·48, 32·3
   (b) any four of 1·06, 2·7, 0·4, 6·48, 32·3

51

1 **(a)** red £9·79   **(b)** red £17·97   **(c)** red £33·96
     blue £9·98   blue £18·69   blue £34·97
     yellow £8·93   yellow £20·87   yellow £25·74
     green £7·92   green £17·99   green £28

2 **(a)** £10·17   **(b)** £24·08   **(c)** £8·02
   (d) £28·05   **(e)** £13·17

3 Alana green   Gita blue
   Jason red   Rosa yellow

52

Pupil's own work.

53

1 **(a)** £2·32   **(b)** £4·12
   (c) £21·13   **(d)** £9·55
   (e) £12·28   **(f)** £25·76

2 **(a)** Swimming Pool £21·48   **(b)** Swimming pool £54·96
       Multi-gym   £26·51   Multi-gym   £61·67
       Running Track   £16·43   Running track   £52·85
       Raquet Sports   £8·14   Raquet sports   £30·94
       Ball Games   £9·91   Ball games   £35·28

3 **(a)** £16·11   **(b)** £57·66

54

1 **(a)** 4·8 kg   **(b)** 6·9 kg   **(c)** 10·8 kg
   (d) 34 kg   **(e)** 7·2 kg   **(f)** 25·9 kg
   (g) 18·4 kg   **(h)** 48·6 kg   **(i)** 20·4 kg

2 **(a)** 7·4   **(b)** 13·6   **(c)** 17·5   **(d)** 15·2
   (e) 21·5   **(f)** 25·2   **(g)** 27·3   **(h)** 20·4
   (i) $3 \times 1·6 = 4·8$ **(j)** $6 \times 1·4 = 8·4$
   (k) $1·1 \times 7 = 7·7$ **(l)** $2·4 \times 4 = 9·6$

3 **(a)** 4·6   **(b)** 3·8   **(c)** 9·2   **(d)** 16·8
   (e) $1·8 \times 2 = 3·6$ **(f)** $2·6 \times 2 = 5·2$
   (g) $2 \times 5·4 = 10·8$ **(h)** $2 \times 7·5 = 15$

4 Team A: 20·9 km
   Team B: 18·7 km

55

1 **(a)** 63·9 kg   **(b)** 28·4 kg   **(c)** 6·96 kg
   (d) 22·59 kg   **(e)** 106·5 kg **(f)** 43·36 kg
   (g) 42·6 kg   **(h)** 15·06 kg **(i)** 170·4 kg

2 **(a)** 87   **(b)** 56·8   **(c)** 140·7   **(d)** 95·4
   (e) 30·85   **(f)** 16·47   **(g)** 21·84   **(h)** 28·56

3 Anna 22·92 ℓ Stuart 10·01 ℓ Alan 14·32 ℓ

4 **(a)** 113·7 ℓ **(b)** 189·5 ℓ **(c)** 151·6 ℓ

5 34·7 ℓ

56

1 (a) 16ℓ (b) 9ℓ (c) 12·5ℓ (d) 8·3ℓ
   (e) 160ℓ (f) 90ℓ (g) 125ℓ (h) 83ℓ

2 (a) 37 (b) 62 (c) 184 (d) 64·1
   (e) 590 (f) 7810 (g) 42 (h) 406
   (i) $10 \times 2·8 = 28$ (j) $0·16 \times 100 = 16$
   (k) $5·3 \times 10 = 53$ (l) $100 \times 1·08 = 108$

3 (a) Pansies 128g Lobelia 74g Marigolds 118g
   (b) Sweet peas 243g Marigolds 177g Cornflowers 216g
   (c) Pansies 448g Sweet peas 567g Marigolds 413g
   (d) Lobelia 333g Sweet peas 729g Cornflowers 648g

4 (a) 128 (b) 85 (c) 504 (d) 444
   (e) 35·7 (f) 13·6 (g) 14·7 (h) 17·5
   (i) 5·4 (j) 688 (k) 412 (l) 714

5 $£42 + £13 = £55$

57

1 (a) 0·7 kg (b) 0·9 kg (c) 1·5 kg

2 (a) 2·4 (b) 0·8 (c) 6·3 (d) 3·1
   (e) $2 \div 10 = 0·2$ (f) $45 \div 10 = 4·5$
   (g) $69 \div 10 = 6·9$ (h) $71 \div 10 = 7·1$

3 (a) 0·92ℓ (b) 0·55ℓ (c) 0·09ℓ

4 (a) 0·17 (b) 0·52 (c) 0·03 (d) 0·66
   (e) $56 \div 100 = 0·56$ (f) $40 \div 100 = 0·4$
   (g) $73 \div 100 = 0·73$ (h) $5 \div 100 = 0·05$

5 (a) $15 \div 10 = 1·5$ (b) $27 \div 100 = 0·27$ (c) $6 \div 100 = 0·06$
   (d) $9 \div 10 = 0·9$ (e) $51 \div 10 = 5·1$ (f) $24 \div 100 = 0·24$

6 (a) 0·24 (b) 0·27 (c) 0·35
   (d) $3 \div 2 = 1·5$ (e) $7·8 \div 2 = 3·9$ (f) $9·2 \div 2 = 4·6$

7 (a) 0·4 kg (b) 0·8ℓ

8 (a) 0·6 (b) 0·7 (c) 0·8 (d) 0·9
   (e) $3·5 \div 5 = 0·7$ (f) $5·4 \div 9 = 0·6$
   (g) $2·8 \div 4 = 0·7$ (h) $5·6 \div 7 = 0·8$

58

1 (a) 3·2 kg (b) 7·8 kg (c) 7·4 kg (d) 9·6 kg

2 (a) 4·3ℓ (b) 3·7ℓ (c) 17·9ℓ (d) 12·3ℓ

3 (a) £1·73 (b) £4·69 (c) £0·19
   (d) £2·47 (e) £0·58 (f) £0·72

4 Cost of chair: $£9·40 \div 4 = £2·35$ and cost of stool: $£7·65 \div 3 = £2·55$
   The chair is cheaper.

59

1 (a) $1\frac{576}{1000}$ kg or 1576 thousandths of 1 kg

   (b) $\frac{449}{1000}$ kg or 449 thousandths of 1 kg

   (c) $9\frac{704}{1000}$ kg or 9704 thousandths of 1 kg

   (d) $\frac{608}{1000}$ kg or 608 thousandths of 1 kg

   (e) $\frac{123}{1000}$ kg or 123 thousandths of 1 kg

   (f) $5\frac{851}{1000}$ kg or 5851 thousandths of 1 kg

(g) $\frac{37}{1000}$ kg or 37 thousandths of I kg

(h) $6\frac{5}{1000}$ kg or 6005 thousandths of I kg

2 (a) 0·982 kg   **(b)** 0·566 kg   **(c)** 0·017 kg   **(d)** 0·309 kg
  (e) 2·294 kg   **(f)** 7·473 kg   **(g)** 4·005 kg   **(h)** 8·708 kg

3 (a) 0·347, 0·348, 0·349, **0·350, 0·351, 0·352**, 0·353
  (b) 1·094, 1·096, 1·098, **1·100, 1·102, 1·104**, 1·106
  (c) 0·716, 0·712, 0·708, **0·704, 0·700, 0·696**, 0·692
  (d) 3·515, 3·510, 3·505, **3·500, 3·495, 3·490**, 3·485
  (e) 0·283, 0·483, 0·683, **0·883, 1·083, 1·283**, 1·483
  (f) 4·199, 4·298, 4·397, **4·496, 4·595, 4·694**, 4·793

4 (a) 0·125   **(b)** 3·818   **(c)** 6·306   **(d)** 5·033

5 (a) Any of 0·998, 0·999, 1·000, 1·001
  (b) Any of 1·232, 1·233, 1·234, 1·235......1·320
  (c) Any of 9·590, 9·591, 9·592, 9·593, 9·594, 9·595, 9·596, 9·597, 9·598, 9·599
  (d) Any of 4·451, 4·452, 4·453, 4·454, 4·455, 4·456, 4·457, 4·458, 4·459

60

1 (a) 9 thousandths   **(b)** 2 tenths   **(c)** 5 hundredths   **(d)** 3 tens
  (e) 0 hundredths   **(f)** 0 units   **(g)** 7 tenths   **(h)** 0 thousandths

2 (a) 0·768   **(b)** 0·687   **(c)** 8·073   **(d)** 40·294
  (e) 6·489   **(f)** 4·902   **(g)** 0·864

3 (a) 0·807   **(b)** 4·269
  (c) 7·031   **(d)** 13·005

4 (a) 7·373   **(b)** 4·080

5 (a) 3·069   **(b)** $5\frac{30}{1000}$

6 (a) 4·543, 4·554, 5·453, 5·534, 5·543
  (b) 8·987, 8·998, 9·897, 9·978, 9·987
  (c) 2·212, 2·201, 2·102, 1·212, 1·21
  (d) 7·65, 7·556, 7·5, 6·765, 6·5

61

1 (a) 5·3 ℓ   **(b)** 8·8 ℓ
  (c) 22·1 ℓ   **(d)** 16·1 ℓ

2 (a) 0·4 ℓ   **(b)** 3·8 ℓ   **(c)** 14·1 ℓ   **(d)** 37·0 ℓ

3

Car	Lap 1	Lap 2
Red	41s	38s
Blue	49s	44s
Orange	39s	40s
Green	52s	55s

4

	Red	Blue	Orange	Green
Lap 1	41	47	36	53
Lap 2	36	42	30	60

(a) Red: 77 s   Blue: 89 s
  Orange: 66 s   Green: 113 s
(b) Red: 5 s   Blue: 5 s
  Orange: 6 s   Green: 7 s

5 (a) Red: 76·9s   Blue: 88·92 s
  Orange: 65·69 s   Green: 112·523 s
(b) Red: 4·5 s   Blue: 5·26 s
  Orange: 6·646 s   Green 6·515 s

62

1 Lorry 12 096 mm, car 4125 mm, bus 9703 mm

2 **(a)** 1842 m   **(b)** 638 ml   **(c)** 7001 g   **(d)** 5287 kg
(e) 3·474 km   **(f)** 0·369 m   **(g)** 2·05 ℓ **(h)** 6·509 kg

3 **(a)** engine oil: 2·8 ℓ screen wash: 3·975 ℓ
(b) engine oil: 1·8 ℓ screen wash: 3·125 ℓ

4 **(a)** 0·1   **(b)** 0·9   **(c)** 0·5   **(d)** 0·2
(e) 0·25   **(f)** 0·75   **(g)** 0·01   **(h)** 0·03
(i) 0·75   **(j)** 0·001   **(k)** 0·467   **(l)** 0·074

5 **(a)** 0·125   **(b)** 0·275   **(c)** 0·1875   **(d)** 0·333….
(e) 0·111…   **(f)** 0·444…

6 **(a)** $\frac{3}{8} = 0·375$, $\frac{2}{5} = 0·4$; $\frac{2}{5}$ is larger

(b) $\frac{9}{10} = 0·9$, $\frac{8}{9} = 0·888….$; $\frac{9}{10}$ is larger

(c) $\frac{2}{7} = 0·285….$, $\frac{1}{3} = 0·333….$; $\frac{1}{3}$ is larger

7 $\frac{5}{8}$, $\frac{13}{20}$, $\frac{2}{3}$, $\frac{27}{40}$

63

1 **(a)** $\frac{15}{100}$, 15%   **(b)** $\frac{96}{100}$, 96%   **(c)** $\frac{10}{50}$ (or $\frac{1}{5}$), 20%   **(d)** $\frac{7}{10}$, 70%

(e) $\frac{3}{6}$ (or $\frac{1}{2}$), 50%   **(f)** $\frac{16}{20}$, 80%   **(g)** $\frac{2}{5}$, 40%   **(h)** $\frac{3}{4}$, 75%

64

1 **(a)** $\frac{31}{100}$, 31%   **(b)** $\frac{30}{60}$ (or $\frac{1}{2}$), 50%   **(c)** $\frac{20}{200}$ (or $\frac{1}{10}$), 10%

(d) $\frac{100}{500}$ (or $\frac{1}{5}$), 20%   **(e)** $\frac{7}{100}$, 7%   **(f)** $\frac{50}{1000}$ (or $\frac{1}{20}$), 5%

2 **(a)** 5%   **(b)** 40%   **(c)** 20%

3 **(a)** 3   **(b)** 20   **(c)** 250
(d) 28   **(e)** 27   **(f)** 28
(g) 16   **(h)** 120   **(i)** 60

4 Quench 480 ml,   Koolora 450 ml,   Sunfresh 490 ml

5 25% is soil

6 40 kg fertiliser
160 kg compost
100 kg sand

65

1 **(a)** 10%   **(b)** 20%   **(c)** 75%
(d) 25%   **(e)** 20%   **(f)** 50%
(g) 40%   **(h)** 5%   **(i)** 34%

2 **(a)** 4   **(b)** 250 m
(c) £75   **(d)** £1·50
(e) 1·2 m   **(f)** 150 stamps
(g) 3·5 kg   **(h)** 10 ℓ

3 **(a)** 4   **(b)** 9   **(c)** 26
(d) 200   **(e)** 3   **(f)** 110

66

1 **(a)** 15%   **(b)** 30%   **(c)** 91%   **(d)** 35%   **(e)** 6%   **(f)** 18%

2 **(a)** 0·43, $\frac{43}{100}$ **(b)** 32%, 0·32   **(c)** 70%, $\frac{7}{10}$

(d) 60% or 0·6   **(e)** 0·08 or $\frac{8}{100}$ **(f)** 12% or $\frac{12}{100}$

3 **(a)** 0·65   **(b)** 27%

4 (a) $\frac{4}{10}$ **(b)** 0·8   **(c)** 2·2   **(d)** 20%

5 (a) T   **(b)** F   **(c)** F   **(d)** T

6 (a) $\frac{3}{4}$, 0·6, 55%   **(b)** 24%, $\frac{4}{20}$

7 9%, 0·9, 99%, 9, 99

8 (a) 0·3, $\frac{35}{100}$, 40%   **(b)** $\frac{14}{100}$, 10%, $\frac{8}{100}$, 0·06

67

1 16 are boys

2 £21

3 200 children

4 12 games

5 Jamie 75%, Roz 74%

6 £109. Most expensive item eligible for $\frac{1}{2}$ price is the sweater (£20).
Use 10% voucher on the jeans, as they are the most expensive item.
Use the £6 off voucher on either the shorts or the gloves.

68

1 (a) 0·7 kg, 700 g   **(b)** 0·1 kg, 100 g
(c) 1·7 kg, 1700 g   **(d)** 2·5 kg, 2500 g

2 A 100 g, 0·1 kg   **E** 300 g, 0·3 kg
B 600 g, 0·6 kg   **F** 900 g, 0·9 kg
C 1000 g, 1 kg   **G** 1300 g, 1·3 kg
D 1600 g, 1·6 kg   **H** 1900 g, 1·9 kg

3 Pupil's own answers.

4 Pupil's own answers.

69

1 Pupil's own answers.

2 Pupil's own answers.

3 (a) 1250 g   **(b)** 3125 g   **(c)** 4568 g   **(d)** 2379 g

4 (a) 5 kg 482 g   **(b)** 7 kg 634 g   **(c)** 9 kg 513 g   **(d)** 6 kg 747 g

5 (a) 7132 g   **(b)** 8 kg 716 g   **(c)** 4303 g   **(d)** 6 kg 948 g
(e) 1255 g   **(f)** 5 kg 681 g   **(g)** 2569 g   **(h)** 9 kg 74 g

70

1 (a) 11 950 kg   **(b)** 34 750 kg   **(c)** 4500 kg

2 (a) 5 tonnes 900 kg   **(b)** 3 tonnes 300 kg   **(c)** 13 tonnes 950 kg

3 Weight of passengers is 3 × 1 tonne = 3 tonnes = 3000 kg
Weight of bus = 8 tonnes 500 kg = 8500 kg
Total weight = 11 500 kg

4

Weight of truck	5 tonnes
4 bags of gravel	2 tonnes
100 paving slabs	2 tonnes
20 sleepers	1 tonne 800 kg
Total weight	10 tonnes 800 kg

5 Practical work.

71

1

	Measure	True length
(a) Firecrest	3 cm	9 cm
(b) Robin	5 cm	15 cm
(c) Goldfinch	4 cm	12 cm
(d) Little Grebe	8 cm	24 cm
(e) Sandpiper	6 cm	18 cm

2 (a)

Wingspan	Measure	True length
Swallow	6 cm	36 cm
Sparrowhawk	7 cm	70 cm
Buzzard	6 cm	120 cm
Golden Eagle	8 cm	240 cm

(b)

Body length	Measure	True length
Swallow	4 cm	24 cm
Sparrowhawk	3·5 cm	35 cm
Buzzard	2 cm	40 cm
Golden Eagle	2·5 cm	75 cm

72

1 Eagles: 30 m by 4 m
Parrots: 8 m by 6 m
Courtyard: 16 m by 8 m
Owls: 14 m by 8 m
Exhibition room: 22 m by 6 m

2 (a) Scots Pine: 30 m
Yew: 9 m
Rowan: 18 m
Spruce: 45 m
Elm: 36 m

(b) Pupil's tree sketches. Heights of sketches should be as follows:
Scots Pine: 10 cm
Yew: 3 cm
Rowan: 6 cm
Spruce: 15 cm
Elm: 12 cm

73

1 (a) 46 mm, 4 cm 6 mm **(b)** 28 mm, 2 cm 8 mm
(c) 53 mm, 5 cm 3 mm **(d)** 72 mm, 7 cm 2 mm
(e) 65 mm, 6 cm 5 mm

2 (a) 37 mm **(b)** 159 mm **(c)** 81 mm **(d)** 104 mm

3 (a) 5 cm 8 mm **(b)** 1 cm 6 mm
(c) 15 cm 3 mm **(d)** 20 cm 5 mm

4 (a) 46 mm, 4 cm 6 mm, 4·6 cm
(b) 51 mm, 5 cm 1 mm, 5·1 cm
(c) 33 mm, 3 cm 3 mm, 3·3 cm

5 Pupil's drawings of caterpillars with lengths of
(a) 26 mm **(b)** 4·3 cm **(c)** 7·4 cm **(d)** 58 mm

74

1. (a) 1 km 400 m + 1 km 700 m = **3100 m**
   (b) $1\frac{1}{2}$ km + 1 km 700 m = **3200 m**
   (c) $2\frac{1}{10}$ km + 1 km 400 m = **3500 m**
   (d) $2\frac{3}{4}$ km + 1 km 700 m = **4450 m**

2. (a) 1 km 650 m + $1\frac{9}{10}$ km + $2\frac{1}{4}$ km = **5800 m**
   (b) $2\frac{1}{4}$ km + $1\frac{1}{2}$ km + 1 km 700 m = **5450 m**
   (c) $2\frac{1}{4}$ km + 1 km 400 m + $2\frac{3}{4}$ km = **6400 m**

3. (a) 4 km 180 m (b) 9 km 40 m (c) 4 km 8 m (d) 5 km 706 m

4. (a) 3200 m and 4 km 300 m
   (b) 4 km 300 m, $3\frac{1}{4}$ km and 3 km 450 m

75

1. belt: 1 m, car: 5 m, calculator: 10 cm, lamppost: 10 m, runway: 1 km, button: 10 mm, Newcastle to York: 100 km, train: 100 m

2. (a) kilometres: distance from the Earth to the Moon, length of a motorway
   (b) metres: length of a bus, height of a tree, height of a house
   (c) cm: length of a pencil
   (d) mm: thickness of a coin, wingspan of a midge

3. Pupil's own suggestions for things to measure in
   (a) kilometres (b) metres
   (c) centimetres (d) millimetres

76

1. (a) 3 m 65 cm
   (b) 3 m 85 cm

2. 5 m 30 cm

3. (a) Dave: 600 mm and 800 mm
      Senga: 250 mm, 350 mm and 600 mm
   (b) 250 mm, 350 mm, 600 mm and 800 mm

4. 22 km 950 m

77

1. (a) 160 mm, 16 cm (b) 165 mm, 16·5 cm
   (c) 170 mm, 17 cm (d) 234 mm, 23·4 cm

2. (a) 82 m (b) 66 m (c) 42 m (d) 46 m (e) 122 m

3. 36 m

78

1. (a) 120 mm (b) 192 mm (c) 138 mm (d) 134 mm

2. To find the perimeter of a rectangle,
   – multiply the sum of the length and breadth by two
   or
   – add double the length to double the breadth.

3. (a) 44 m (b) 52 m (c) 50 m (d) 62 m

4. (a) 110 m (b) 134 m (c) 90 m

5. 250 m

79

1 (a) 07:00 (b) 08:00 (c) 08:55
(d) 10:13 (e) 11:32 (f) 12:00
(g) 14:07 (h) 16:43 (i) 17:00

2 (a) In bed (b) In a taxi (c) About to board a plane
(d) On a plane (e) On a plane on the runway

3 (a) 4.00am (b) 10.00pm (c) 2.15am
(d) 7.45pm (e) 11.09pm (f) 12.23am
(g) 8.36pm (h) 6.51pm (i) 1.11am
(j) 12.00 midnight

4 (a) Edinburgh (b) Bristol (c) Aberdeen (d) Belfast
(e) Glasgow (f) Manchester (g) Newcastle (h) Dublin

5

Departure	Time
Edinburgh	05:00
Belfast	06:01
Newcastle	09:35
Dublin	12:27
Aberdeen	13:27
Manchester	17:44
Glasgow	20:11
Bristol	21:35

80

1 London is 5 hours ahead of New York.
London is 3 hours ahead of Bermuda.
London is 2 hours behind Athens.
London is 5 hours 30 minutes behind Bombay.
London is 8 hours behind Beijing.
London is 9 hours 30 minutes behind Adelaide.

2 (a) 16:30 (b) 20:00 (c) 11:30

3 (a) 15:00 (b) 08:00 (c) 11:00

4 (a) 10:00 (b) 05:00 (c) 13:30

5 (a) 11:30
(b) Cyprus   15:45
Moscow   17:15
Washington   17:00
San Francisco 21:00

81

1 (a) 11:05 (b) 13:25 (c) 18:40
(d) 23:15 (e) 08:25 (f) 00:35

2 (a) 06:50 (b) 09:20 (c) 17:00
(d) 10:25 (e) 17:25 (f) 23:40

3 (a) 1 hour 15 minutes (d) 1 hour 50 minutes
(b) 1 hour 30 minutes (e) 1 hour 45 minutes
(c) 1 hour 40 minutes (f) 2 hours 55 minutes

82

1 1 hour 25 minutes

2 10:30

3 11:35

4 13:05

5 1 hour 45 minutes

6 13:30

7 5 hours 35 minutes

1 11:35 train

2 55 minutes

3 **(a)** 3 hours 10 minutes   **(b)** 19:45

4 Yes. Take the 07:10 train from Ambertini to Cala, then the 09:05 boat from Cala to Bertoli, then the 10:10 boat from Bertoli to Rivera. Then take the 14:05 train from Rivera to Ambertini.

5 Pupil's own work.

1

weight cooking time	1 kg	$1\frac{1}{4}$ kg	$1\frac{1}{2}$ kg	$1\frac{3}{4}$ kg	2 kg	$2\frac{1}{4}$ kg	$2\frac{1}{2}$ kg
Chicken	40 min	50 min	60 min	70 min	80 min	90 min	100 min
Lamb	48 min	60 min	72 min	84 min	96 min	108 min	120 min
Beef	60 min	75 min	90 min	105 min	120 min	135 min	150 min

2 **(a)** 11.40am   **(b)** 10.30am   **(c)** 11.36am

3 **(a)** 3 min

(b) 3 hours

(c) $1\frac{1}{2}$ hours

4 12 hours 30 minutes

5 **(a)** Yes, she has lived for 1100 weeks approximately

(b) No, she has lived for 7665 days approximately

(c) Yes, she has lived for 183 960 hours approximately

1 **(a)** 20 seconds   **(b)** 28 seconds   **(c)** 42 seconds
(d) 44 seconds   **(e)** 50 seconds   **(f)** 59 seconds

2 **(a)** 10 minutes 5 seconds   **(b)** 17 minutes 12 seconds
(c) 23 minutes 30 seconds   **(d)** 30 minutes 23 seconds
(e) 4 minutes 38 seconds   **(f)** 13 minutes 46 seconds

1 Pupil's own work.

2 Pupil's own work.

1 **(a)** 2800 m²
(b) 2000 m²
(c) 900 m²
(d) 5700 m²

2 9600 m²

3 **(a)** 14 cm²
(b) 19 cm²
(c) 33 cm²

1 **(a)** 9 m²   **(b)** 14 m²   **(c)** 12·5 m²   **(d)** 8 m²
(e) 12 m²   **(f)** 4·5 m²   **(g)** 17·5 m²

2 (a) 6 cm²

(b) 5 cm²

(c) 7·5 cm²

(d) 7 cm²

3 (a) 36·5 m²

(b) 44 m²

89

1 32 slabs

2 2 cans. Fence is 36m² sq, each can covers 20 m²

3 36 cm²

4 50 m²

5 48 × 1, 24 × 2, 16 × 3, 12 × 4, 8 × 6

6 12 m²

90

1 (a) 700 ml **(b)** 450 ml **(c)** 750 ml **(d)** 200 ml **(e)** 1090 ml

(f) 80 cl **(g)** 50 cl **(h)** 40 cl **(i)** 25 cl **(j)** 7 cl

2 (a) 33 cl **(b)** $\frac{1}{10}\ell$

(c) 80 cl **(d)** 5 cl

3 (a) 6 ℓ, 66 cl, 606 ml, $\frac{6}{10}\ell$, 6 cl

(b) 1003 ml, 1ℓ 3 cl, 1·3 ℓ, 133 cl, 13 ℓ

4 400 ml, $\frac{1}{2}\ell$, 0·6 ℓ

91

1 milk: **Y** water: **Z** vinegar: **X** olive oil: **W**

2 Pupil's own work. Typical capacity for type of containers:

A 40 cl **B** 75 cl **C** 25 cl **D** 35 cl **E** 100 cl **F** 15 cl

3 Pupil's own work.

92

1 (a) Jack: T Mark: F Sophie: T Zoe: F

(b) Rosie: 875 ml Leela: 1ℓ

2 (a) 1060 ml **(b)** 1100 ml

(c) 1900 ml **(d)** 1810 ml

(e) 2150 ml **(f)** 1760 ml

3 130 ml

4 (a) 35 cl and 550 ml **or** $\frac{4}{10}\ell$ and 50 cl **(b)** $\frac{4}{10}\ell$ and 35 cl

(c) 45 cl and 35 cl **(d)** 45 cl and 550 ml

93

1 300 g

2 200 g

3 123 ml

4 204 ml

5 10 tablespoons

6 35 minutes

7 12:10pm

8 No. The cake will be 15 days old.

94

1 (a) $3\frac{1}{2}$ pints (b) 175 pints (c) 21 pints (d) 8 pints

2 (a) 45 ℓ (b) 112·5 ℓ (c) 675 ℓ (d) 2250 ℓ

3 (a) 120 g (b) 210 g (c) 360 g (d) 315 g

4 1 kg

95

1 (a) 24 km (b) 64 km (c) 96 km (d) 44 km

2 (a) Crater (b) Zircon Mine

3 (a) 22 pounds (b) 17·6 pounds (c) 44 pounds (d) 34·1 pounds

4 (a) pints / centilitres (b) ounces / grams (c) miles / kilometres
   (d) gallons / litres (e) pounds / kilograms

96

1 Pupil's own designs.

2 Pupil's own designs.

97

1 (a) Pupil's own work.
   (b) Pupil's own work.
   (c) 4, 2, 1, 3
   (d) Pupil's own work.

2 (a) Pupil's own work.

(b)

Regular polygon	Number of equal sides and equal angles	Number of times it fits its outline
Equilateral triangle	3	3
Square	4	4
Regular polygon	5	5
Regular hexagon	6	6
Regular heptagon	7	7
Regular octagon	8	8

(c) The number of equal sides/angles is the same as the number of times it fits its outline.

98

1 (a) A B F G I J K (b) C E F G H (c) K L
   (d) L (e) B F G I (f) E (g) K
   (h) A (i) K (j) B F G I

2 **A** kite **B** parallelogram **C** equilateral triangle **D** isosceles triangle
   E regular octagon **F** rhombus **G** square **H** regular pentagon
   I rectangle **J** quadrilateral **K** trapezium **L** hexagon **J** quadrilateral

99

1 (a) D (b) B C F I (c) A H J (d) E G

2 (a) A E G J (b) A (c) G (d) E
   (e) I (f) C (g) I (h) I
   (i) F (j) C (k) D

3 **A** rectangle **B** isosceles trapezium **C** isosceles triangle
   D right-angled scalene triangle **E** equilateral triangle **F** kite
   G regular hexagon **H** rhombus **I** pentagon **J** parallelogram

4 Pupil's own work.

100

1 Pupil's own work.

2 Pupil's own work.

101

1 Pupil's own work.

102

1 Pupil's own work.
2 Pupil's own work.

103

1 **(a)** Yes   **(b)** Yes   **(c)** No   **(d)** No
(e) Yes   **(f)** Yes   **(g)** Yes   **(h)** No

2 Pupil's own work.

104

1 **(a)** $96 \, cm^2$ **(b)** $370 \, cm^2$ **(c)** $149 \, cm^2$
(d) $268 \, cm^2$ **(e)** $84 \, cm^2$ **(f)** $144 \, cm^2$

2 **(a)** cuboid   **(b)** $78 \, cm^2$

3 3 cm

105

1 **(a)** 11 cubes   **(b)** 12 cubes   **(c)** 13 cubes   **(d)** 23 cubes

2 **(a)** 24 cubes   **(b)** 13 cubes

3 **(a)** 4 cubes   **(b)** 5 cubes

106

1 **(a)** $(4, {}^-3)$ **(b)** $(6, 1)$ **(c)** $(5, 0)$ **(d)** $({}^-3, 0)$
(e) $(0, 3)$ **(f)** $(0, {}^-1)$ **(g)** $({}^-2, {}^-2)$

2 **(a)** black   **(b)** red   **(c)** orange   **(d)** blue
(e) brown   **(f)** white   **(g)** green   **(h)** yellow

107

1 • $(3, 0), ({}^-1, 0), ({}^-2, 1), ({}^-6, 1), ({}^-7, 2), ({}^-6, 3), ({}^-2, 3), ({}^-1, 2), (3, 2), (4, 1),$

 • $(3, 0), ({}^-1, 0), ({}^-2, {}^-1), ({}^-6, {}^-1), ({}^-7, {}^-2), ({}^-6, {}^-3), ({}^-2, {}^-3), ({}^-1, {}^-2), (3, {}^-2), (4, {}^-1)$

2 • $({}^-2, 1), ({}^-6, 1), ({}^-6, 2), ({}^-5, 2), ({}^-5, 4) ({}^-1, 4), ({}^-1, 3), ({}^-2, 3)$
horizontal reflection:
$({}^-2, {}^-1), ({}^-6, {}^-1), ({}^-6, {}^-2), ({}^-5, {}^-2)$
$({}^-5, {}^-4), ({}^-1, {}^-4), ({}^-1, {}^-3), ({}^-2, {}^-3)$
vertical reflection:
$(2, 1), (6, 1), (6, 2), (5, 2)$
$(5, 4), (1, 4), (1, 3), (2, 3)$

108

1 **(a)** In the position shown: $({}^-1, 2), ({}^-6, 1), ({}^-4, 3)$
   After translation: $(5, 2), (0, 1), (2, 3)$

   (b) In the position shown: $(1, {}^-1), (6, {}^-1), (4, {}^-3), (3, {}^-3)$
   After translation: $({}^-4, {}^-1), (1, {}^-1), ({}^-1, {}^-3), ({}^-2, {}^-3)$

   (c) In the position shown: $({}^-3, {}^-1), ({}^-2, {}^-2), ({}^-3, {}^-3), ({}^-7, {}^-2)$
   After translation: $(5, 0), (6, {}^-1), (5, {}^-2), (1, {}^-1)$

   (d) In the position shown: $(0, 2), (1, 3), (5, 1), (1, 1)$
   After translation: $({}^-6, 0), ({}^-5, 1), ({}^-1, {}^-1), ({}^-5, {}^-1)$

(e) In the position shown: (⁻4, 0), (⁻7, 0), (⁻6, 2), (⁻4, 2)
After translation: (1, ⁻2), (⁻2, ⁻2), (⁻1, 0), (1, 0)

(f) In the position shown: (5, 0), (7, 0), (5, ⁻2), (3, ⁻2)
After translation: (⁻2, 2), (0, 2), (⁻2, 0), (⁻4, 0)

109

1 (a) In the position shown: (0, 0), (0,3), (2, 0)
After rotation: (0,0), (⁻3, 0), (0, 2)

(b) In the position shown: (0, 0), (⁻1, 0), (⁻3, 2), (⁻2,3), (0,1)
After rotation: (0, 0), (1, 0), (3, ⁻2), (2, ⁻3), (0, ⁻1)

(c) In the position shown: (0, 0), (⁻2, 0), (⁻3, 1), (⁻2, 2), (0, 2), (⁻1, 1)
After rotation: (0, 0), (0, 2), (1, 3), (2, 2), (2, 0), (1, 1)

(d) In the position shown: (0, 0), (0, 2), (1, 2), (1, 3), (3, 3), (3, 1), (2, 1), (2, 0)
After rotation: (0, 0), (0, ⁻2,), (⁻1, ⁻2), (⁻1, ⁻3), (⁻3, ⁻3), (⁻3, ⁻1), (⁻2, ⁻1), (⁻2, 0,)

110

1 • acute: **I, L** • obtuse: **J, M** • reflex: **K, N**

2 P 80°, Q 35°, R 110°, S 155°, T 47°, U 122°

3 Check Pupil's drawings to show angles of
(a) 60° **(b)** 145° **(c)** 28° **(d)** 99°

111

1 V 320° W 95° X 278° Y 171°

2 A 60° B 100° C 90° D 75° E 59° F 27°

112

1 (a) Saturday, last week **(b)** Thursday, this week
(c) Wednesday, last week **(d)** Sunday, this week

2 (a) 9 **(b)** 13

3 The graph shows that busiest times are at weekends and the least busiest times are midweek.

4 (a)

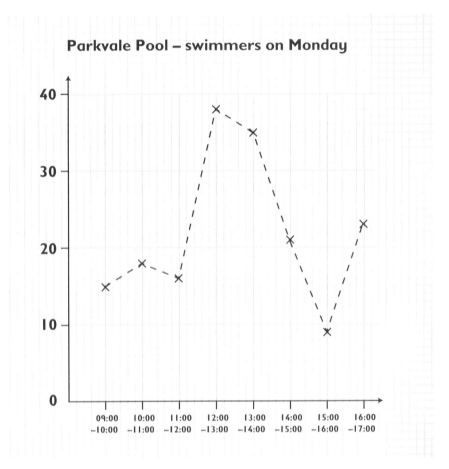

Parkvale Pool – swimmers on Monday

(b) The trend graph shows that it is busiest from noon until mid-afternoon; the number of swimmers is lower but about the same for each hour in the morning.

113

1 (a) Smallest: Alan. Greatest: Luke. Range: 12

(b) Bindu and Dave; mode 51

(c) Sammy; median 55

2 (a) 50, 54, 55, 58, 60, 60, 62

(b) range 12; mode 60; median 58

3 (a) 49, 50, 52, 52, 54, 57, 58, 60, 60, 60, 64

(b) range 15; mode 60; median 57

4 (a) 48, 49, 50, 50, 51, 51, 52, 52, 54, 54, 55, 55, 56, 57, 57, 58, 58, 60, 60, 60, 60, 60, 60, 62, 64

(b) range 16; mode 60; median 56

114

1 (a) Bindu   **(b)** Sammy

2 (a) 45   **(b)** 80   **(c)** 90

3 (a) 15   **(b)** 30   **(c)** 27

4 Red team: range 9;   mode 18;   median 21
Yellow team: range 15;   mode 24;   median 21
Orange team: range 18;   mode 12;   median 12

115

1 (a) Anne 20;   Sammy 17

(b) 126   **(c)** 18

(d) Less than the mean: Dave, Sammy, Dion
Greater than the mean: Bindu, Anne
Equal to the mean: Alan, Luke

2 (a) 133   **(b)** 19

(c) Above the mean: Steve, Lisa, Dianne, Sheena
Below the mean: Paul, Nazir, Joan

3 (a) Range 7; mode 18; median 19

(b) 20

(c) Above the mean: 4; below the mean: 6; equal to the mean: 1

116

1 (a) £36

(b) Adele £9,   Ruth £7,   Gabby £5,   Zena £8,   Carla £7

(c) $\frac{9}{36}$ $(\frac{1}{4})$

(d)

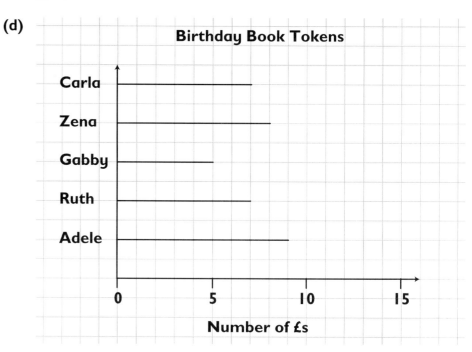

(a) £30   (b) Omar   (c) Adam and John

(d) Omar £10:   Adam £5;   Tony £6;   John £5;   Feng £4

3 (a) graph A   (b) graph A   (c) graph B   (d) graph B

117

Red £56;   Yellow £60;   Blue £40·75;   Green £45·50

118

1 (a) See answers to Pupil Sheet 57.

(b) Most money: Yellow group. Least money: Blue group.

(c) Red £11·20;   Yellow £12;   Blue £8·15;   Green £9·10.

2 (a) See answers to Pupil Sheet 57.

(b) Red £10·80;   Blue £9·75

3 See answers to Pupil Sheet 57.

119/120

1 (a) Quad, Venta   (b) Svensun 5, Sherwood, Norlund

(c) Morten, Venta   (d) Morten, Rodex

(e) Sherwood, Svensun 12   (f) Sherwood, Norlund

2 Hands free

3 (a) Svensun 12   (b) Norlund   (c) Sherwood

4 Pupil's own work.

121

1 (a) 4   (b) greatest: 140–149;   smallest: 130–139

(c) shortest: 130–139; tallest: 160–169

(d) shorter than 140 cm: 14; between 149 cm and 160 cm: 22

(e) taller $\frac{1}{5}$;   between $\frac{7}{20}$

(f) It is true that five eighths of the children in Year 6 are between 139 and 160 centimetres tall as there are 50 children out of a total of 80 who fall between these two heights.

2 Pupil's own work.

122

1 (a) likely   (b) impossible   (c) certain   (d) unlikely

2 Children's answers will vary.

3 Children's answers will vary.

123

1 (a) 3, 7, 11, 21

(b) 4, 12, 16, 20

(c) 4, 16

(d) 3, 7, 11

2 (a) 2, 4, 6, 8   (b) 3, 6, 9   (c) 2, 3, 4, 6   (d) 3, 6

3 (a) $\frac{1}{6}$ (b) $\frac{1}{2}$ (c) $\frac{2}{3}$ (d) 1

4 (a) Less than one in two.

(b) One in two.

E1

1 (a) 17 (b) 236 (c) 1022 (d) 2359 (e) 3503

2 (a) ∩|||| (14) (b) ∩∩∩∩|||| (48)
 ||||

 (c) ꝯ∩∩∩∩| (161) (d) ꝯꝯꝯ∩∩∩∩∩||| (2085)
 ꝯꝯ∩∩∩||

3 (a) Pupil's own answers.

 (b) ꝯꝯꝯ∩∩∩||| (366)
 |||

 (c) Children's answers will vary.

4 (a) stones: 379 logs: 3728

 (b) stones: 305 logs: 1222

E2

1 (a) £15 863 178 (b) £5 444 302

2 (a) Total: £13 194 427 Difference: £5 946 163

 (b) Total: £14 707 637 Difference: £2 501 763

 (c) Total: £7 762 657 Difference: £7 224 499

 (d) Total: £20 072 371 Difference: £1 743 669

3 (a) £14 856 837 (b) £3 790 347

E3

1 (a) One month: £9456 one year: £113 472

 (b) One month: £7884 one year: £94 608

 (c) One month: £7936 one year: £95 232

 (d) One month: £7700 one year: £92 400

 (e) One month: £6573 one year: £78 876

2 (a) Option Two [Option 1: £302 808 Option 2: £318 948]

 (b) £16 140

E4

1 (a) 101 (b) 110 (c) 130 (d) 106

2 (a) 152 (b) 161 (c) 201

3 (a) 220 (b) 205 (c) 250 (d) 310

E5

1 (a) £3272, £6536, £20 160, £9824, £8352, £5488, £41 792, £17 616

 (b)

Prices (£s) exactly divisible by 8	3272	6536	20 160	9824	8352	5488	41 792	17 616
Last 3 digits of the price	272	536	160	824	352	488	792	616
Are the last 3 digits exactly divisible by 8?	Yes	Yes	Yes	Yes	Yes	Yes	Yes	Yes

 (c) A number is exactly divisible by 8 if the last 3 digits are exactly divisible by 8.

2 (a) £9855, £6804, £87 246, £1422, £2997, £7362, £34 731, £969 318

 (b)

Prices (£s) exactly divisible by 9	9855	6804	87 246	1422	2997	7362	34 731	969 318
Sum of the digits in the price	27	18	27	9	27	18	18	36
Is the digit sum exactly divisible by 9?	Yes	Yes	Yes	Yes	Yes	Yes	Yes	Yes

(c) A number is exactly divisible by 9 if the sum of its digits is a multiple of 9.

3 Exactly divisible by 8: 8856, 24 632, 59 904, 7424

Not exactly divisible by 9: 24 632, 32 178, 7424

Numbers with > 4 digits have their digits grouped in 3s from the right.

e.g 1 234 567

E6

1 (a)

Number of triangles	Number of candles
1	→ 3
2	→ 6
3	→ 9
4	→ 12
5	→ 15

(b) 10 triangles: 30 candles; 100 triangles: 300 candles

(c) The number of candles is **3** times the number of triangles.

2 (a)

Number of rhombuses	Number of candles
1	→ 4
2	→ 8
3	→ 12
4	→ 16
5	→ 20

(b) 10 rhombuses: 40 candles; 50 rhombuses: 200 candles

(c) The number of candles is **4** times the number of rhombuses.

(d) 32 candles: 8 rhombuses; 80 candles: 20 rhombuses

E7

1 (a)

Number of squares	Number of candles
1	→ 4
2	→ 8
3	→ 12
4	→ 16
5	→ 20
8	→ 32
11	→ 44

(b) The number of candles is 4 times the number of squares.

The number of squares is the number of candles divided by 4.

2 (a)

Number of squares	Number of drops
1	→ 0
2	→ 1
3	→ 2
4	→ 3
6	→ 5
9	→ 8
15	→ 14

(b) The number of drops is the number of squares subtract 1.

The number of squares is the number of drops add 1.

(c) 19 drops: 20 squares; 99 drops: 100 squares.

3 (a)

Number of triangles	Number of drops
1	→ 3
2	→ 4
3	→ 5
4	→ 6
7	→ 9
10	→ 12
21	→ 23

(b) The number of drops is the number of triangles add 2.
   The number of triangles is the number of drops subtract 2.

(c) 52 drops: 50 triangles   101 drops: 99 triangles

E8

1 15 hens

2 3 horses

3 4 goats

4 (a) 18   (b) 15   (c) 9   (d) 6

5 (a) 9 brown, 6 white   (b) 15 brown, 10 white
   (c) 30 brown, 20 white   (d) 18 brown, 12 white

E9

1 (a) 8 long-haired   (b) 4 not long-haired

2 (a) 8 male   (b) 6 female

3 (a) 18 kg   (b) 6 kg

4 (a) 12 tins   (b) 16 tins

5 (a) 1 horse to 2 pigs
   (b) 2 sheep to 1 cow
   (c) 1 horse to 3 goats
   (d) 2 cows to 3 pigs
   (e) 1 horse in every 10 animals
   (f) 1 pig in every 5 animals

E10

1 (a) £10·72
   (b) The multipack is better value, with each tin costing £1·27 a tin; this
      is 7p cheaper than buying tins singly.

2 (a) 43p   (b) 46p
   The family pack is better value as it is 3p cheaper per tin.

3 (a) The triple pack is better value at £2·05 a tin compared to £2·14 a
      tin for the economy pack, a saving of 9p per tin.
   (b) The 7 kg pack is better value at £1·30 per kg compared to £1·35
      per kg for the 5 kg pack, a saving of 5p per kg.

4 (a) £7·65   (b) £7·20   (c) £7·90

5 (a) £3·85   (b) 77p   (c) £1·52   (d) 38p

E11

1 (a) Jack 0·939 seconds;   Zoë 1·177 seconds;
      Gemma 1·565 seconds;   Pete 1·578 seconds.
   (b) Round 1: Red team 0·958 seconds; Yellow team 1·354 seconds.
      Round 2: Red team 1·158 seconds; Yellow team 1·789 seconds.

2 (a) Round 1: Red team 7·995 seconds; Yellow team 5·478 seconds.
      Round 2: Red team 7·777 seconds; Yellow team 7·86 seconds.
   (b) Jack 6·486 seconds; Zoë 9·286 seconds; Gemma 5·989 seconds;
      Pete 7·349 seconds.

3 Total times for both events:

Jack 7·425 seconds   Zoë 10·463 seconds

Gemma 7·554 seconds   Pete 8·927 seconds

4 Jack and Gemma reached the finals.

E12

(Game)

E13

1 Robot Zone: 2600 m^2 Digital Zone: 2000 m^2

Micro Zone: 1500 m^2 Internet Zone: 2600 m^2

2 **(a)** 16 cm^2 **(b)** 15 cm^2 **(c)** 36 cm^2

E14

1 Daylight   08.07

Walkabout   11.34

House Call   14.19

Music Chart   20.31

Nighthawk   23.04

Motorsport   00.45

2 Fan Club   06.34

Seascape   08.35

Amazon Trek   12.33

Gardening Life   17.41

Sportsweek   23.53

3 **(a)** 68 min   **(b)** 72 min   **(c)** 87 min

(d) 104 min   **(e)** 109 min   **(f)** 118 min

E15

1 06.56

2 07.18

3 **(a)** 54 minutes   **(b)** 11.34

4 **(a)** 19 minutes

(b) 46 minutes

5 **(a)** 16.16   **(b)** 9 hours 57 minutes

E16

1 Cuboid

	Number of cubes in one row	(length) Number of rows	(breadth) Number of layers	(height) Volume in centimetre cubes
yellow	6	4	3	72 cm^3
red	7	5	2	70 cm^3
green	8	6	4	192 cm^3
purple	7	4	4	112 cm^3
blue	9	5	3	135 cm^3

2 Pupil's own answers; e.g.: To find the volume mulitply the length by the breadth by the height.

3 yellow 240 cm^2 red 210 cm^2 orange 960 cm^2

brown 448 cm^2 blue 1000 cm^2 green 450 cm^2.

E17

Pupil's own answers (must have rotational symmetry)

E18

Pupil's own work.

E19

Pupil's own work.

E20

Pupil's own work.

E21

1 (a) 3 points: 3 intersections 4 points: 6 intersections 5 points:
10 intersections

(b)
Points lines drawn from	2 2 2 2
Points lines drawn to	2 3 4 5
Number of intersections	1 3 6 10

2 (a) 15   (b) 21   (c) 28   (d) 36   (e) 45

E22

1 (a) Vertices of half shape: (⁻3, 1), (⁻1, 3), (0, 2), (2, 2), (3, 3), (3, 1).
Vertices needed to complete whole shape (where red line is an axis of symmetry): (⁻1, ⁻1), (0, 0), (2, 0), (3, ⁻1).

(b) Vertices of half shape: (⁻2, ⁻3), (⁻3, ⁻3), (⁻3, ⁻2), (⁻5, 0), (⁻5, 1), (⁻3, 3), (⁻2, 3).
Vertices needed to complete whole shape (where red line is an axis of symmetry): (⁻1, ⁻3), (⁻1, ⁻2), (1, 0), (1, 1), (⁻1, 3).

(c) Vertices of half shape: (⁻2, 2), (2, 2), (2, 1), (3, 1), (2, 0), (3, ⁻1), (2, ⁻2).
Vertices needed to complete whole shape (where red line is an axis of symmetry): (0, ⁻2), (⁻1, ⁻3), (⁻1, ⁻2), (⁻2, ⁻2), (1, ⁻3).

2 Vertices of the shape: (1, ⁻1), (2, ⁻1), (2, ⁻2), (3, ⁻2), (3, ⁻1), (4, ⁻1), (4, ⁻3), (1, ⁻3).
Vertices of the shape's reflection when both red lines are lines of symmetry: (1, ⁻1), (0, ⁻1), (0, 0), (⁻1, 0), (⁻1, ⁻1), (⁻2, ⁻1), (⁻2, 1), (1, 1).

E23

1 (a) The greatest number of children: written method.
The smallest number of children: mental method.

(b) written: $\frac{1}{2}$ mental: $\frac{1}{8}$ calculator: $\frac{3}{8}$

2 (a) True   (b) False   (c) False   (d) True

3 (a) 10

(b) Number: 1, Measure: 3, Shape: 4, Data handling: 2

(c) Number: 10% Measure: 30%

(d) Shape: $\frac{2}{5}$ Data handling: $\frac{1}{5}$

(e) Data handling and Measure were chosen by half of the children;
Shape and Number were also chosen by half the children.

E24

1 (a) Anne and Luke

(b) 24 km

(c) Dion

(d) Bindu 7 km; Luke 4 km; Dave 3 km; Anne 4 km; Dion 6 km.

(e) Dion and Bindu

2 (a) Bar-line chart   (b) Compound bar chart   (c) Pie chart
(d) Bar-line chart   (e) Pie chart

3 (a)

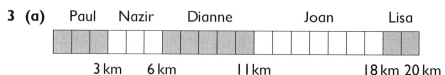

Paul   Nazir   Dianne   Joan   Lisa

3 km   6 km   11 km   18 km 20 km

(b)

Bar-line chart

Distance
in km

Paul Nazir Dianne Joan Lisa

4 £30

1 **(a)** 10 cents **(b)** 10 pence

2 **(a)** €1·60 **(b)** €5·60 **(c)** €3·60
(d) £2·50 **(e)** £1·50 **(f)** £3·75

3

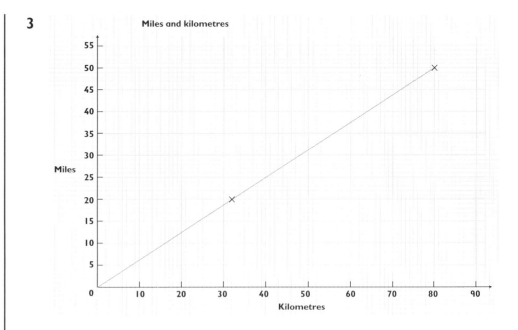

4 **(a)** 25 miles **(b)** 15 miles **(c)** 37·5 miles **(d)** 8 km
(e) 48 km **(f)** 76 km

1 Pupil's own results.

2 Pupil's own results.

3 Pupil's own results.

1 Pupil's own results. Their results should show that the die is unfair.

2 Pupil's own results. Their results should show that the die is unfair.

3 Pupil's own results. Their results should show that the spinner is unfair.

E28

1 Population in Year 2000 is 1 024 000

2 **(a)** 100 red tokens   **(b)** 200 blue tokens

3 720

4 45

5 After 29 more days, as 30 is the next multiple of 6, 15 and 10.

E29

1 The 9 / 12 tile.

2 Rule for making 11: Multiply 5 and 3, multiply 4 and 1 then subtract.
Rule for making 13: Multiply 6 and 8, multiply 7 and 5 then subtract the product of 7 and 5 (35) from the product of 6 and 8 (48).
Rule for making 10: Multiply 7 and 4, multiply 6 and 3 then subtract the product of 6 and 3 (18) from the product of 7 and 4 (28).
Rule for making 30: Multiply 9 and 8, multiply 6 and 7 then subtract the product of 6 and 7 (42) from the product of 9 and 8 (72).
Rule for making 31: Multiply 9 and 7, multiply 8 and 4 then subtract the product of 4 and 8 (32) from the product of 9 and 7 (63)

3 There are 85 tricks in the magician's box.

4 **(a)** 5   **(b)** 10

E30

1 **(a)** Gold: £205·73   Silver: £3·08   Platinum: £317·10
Rule for making 11: Multiply 5 and 3, multiply 4 and 1. Subtract the products, 15 − 4 = 11

  (b)

10 oz gold	2057·30
9 oz silver	27·72
12 oz platinum	3805·20
	5890·22

  (c)

5 oz platinum	£1585·50
11 oz gold	£2263·03
Difference	£677·53

2 **(a)** £1642·50   **(b)** £6366·60   **(c)** £15 185·25

3 Gold: $264·51   Platinum: $407·70

4 **(a)** 603 g   **(b)** £21 470 117

Pupil Sheet 5

Name:

421 + 387 → (400 + 21) + (400 − 13) → 800 + 21 − 13 = 808

214 + 197 → (200 + 14) + (200 − 3) → 400 + 14 − 3 = 411

296 + 308 → (300 − 4) + (300 + 8) → 600 − 4 + 8 = 604

518 + 485 → (500 + 18) + (500 − 15) → 1000 + 18 − 15 = 1003

319 + 295 → 600 + 19 − 5 = 614

696 + 716 → 1400 − 4 + 16 = 1412

213 + 178 → 400 + 13 − 22 = 391

825 + 789 → 1600 + 25 − 11 = 1614

423 + 377 → 800 + 23 − 23 = 800

586 + 615 → 1200 − 14 + 15 = 1201

Addition • Near multiples of 100, doubling strategy

Pupil Sheet 7

Name:

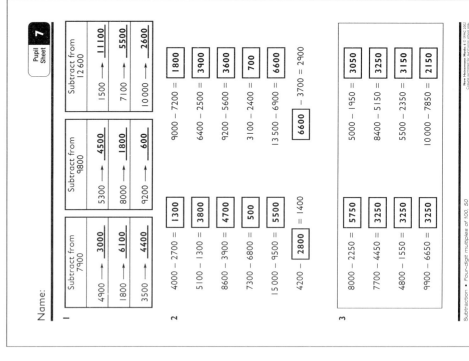

1

Subtract from 7900	Subtract from 9800	Subtract from 12 600
4900 → 3000	5300 → 4500	1500 → 11 100
1800 → 6100	8000 → 1800	7100 → 5500
3500 → 4400	9200 → 600	10 000 → 2600

2

4000 − 2700 = 1300   9000 − 7200 = 1800

5100 − 1300 = 3800   6400 − 2500 = 3900

8600 − 3900 = 4700   9200 − 5600 = 3600

7300 − 6800 = 500   3100 − 2400 = 700

15 000 − 9500 = 5500   13 500 − 6900 = 6600

4200 − 2800 = 1400   6600 − 3700 = 2900

3

8000 − 2250 = 5750   5000 − 1950 = 3050

7700 − 4450 = 3250   8400 − 5150 = 3250

4800 − 1550 = 3250   5500 − 2350 = 3150

9900 − 6650 = 3250   10 000 − 7850 = 2150

Subtraction • Four-digit multiples of 100, 50

Pupil Sheet 4

Name:

1 Estimate the numbers shown by the arrows on each line.

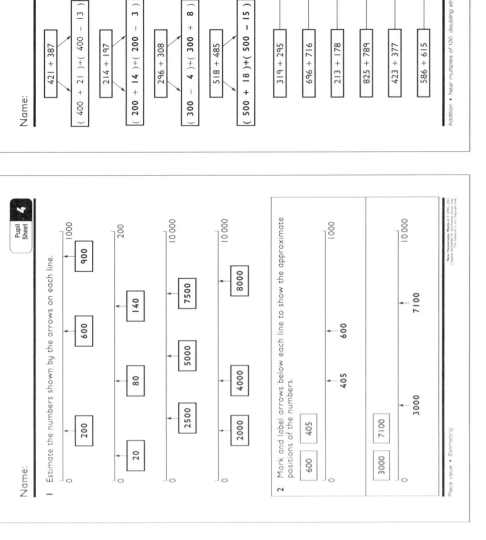

0 ———— 200   20   80   140

0 ———— 1000   200   600   900

0 ———— 10 000   2000   2500   4000   5000   7500

0 ———— 10 000   8000

2 Mark and label arrows below each line to show the approximate positions of the numbers.

0 ———— 1000   600   405

0 ———— 1000   600   405

0 ———— 10 000   3000   7100

0 ———— 10 000   3000   7100

Place value • Estimating

Pupil Sheet 6

Name:

1 What is the total amount to be paid to

• Brookes and Quikfix   £8800   • Quikfix and Minvac   £8000

• Scotvale and Complus   £5200   • Minvac and Afkar?   £7700

Pay the sum of: ____   to: ____   (× 6 boxes)

2

6100 + 2300 = 8400   2700 + 5300 = 8000

4400 + 4800 = 9200   1900 + 6500 = 8400

1500 + 5800 = 7300   5600 + 3600 = 9200

3 How much money altogether is in cash boxes

A	B	C	D	E	F
£8700	£9100	£3600	£2800	£5400	£7900

• B and E   £14 500   • F and B   £17 000   • D and E   £8200

• A and D   £11 500   • C and F   £11 500   • C and A?   £12 300

4

6500 + 7400 = 13 900   8400 + 6700 = 15 100

5300 + 5800 = 11 100   4900 + 7600 = 12 500

3300 + 5900 = 9900   7600 + 5800 = 13 400

3300 + 5800 = 13 200

Addition • Four-digit multiples of 100

Pupil Sheet 8

Name:

1 8000 − 2785

2785 + 5 + 10 + 200 + 5000 = 8000

8000 − 2785 = **5215**

6000 − 3471

3471 + 9 + 20 + 500 + 2000 = 6000

6000 − 3471 = **2529**

9000 − 5163

5163 + 7 + 30 + 800 + 3000 = 9000

9000 − 5163 = **3837**

10 000 − 4232

4232 + 8 + 60 + 700 + 5000 = 10 000

10 000 − 4232 = **5768**

2 7000 − 1546 = **5454**   5000 − 2807 = **2193**

8000 − 4320 = **3680**   4000 − 1094 = **2906**

9000 − 3333 = **5667**   3000 − 888 = **2112**

3 Find the difference between

1658 and 5000 **3342**   10 000 and 8911 **1089**   2079 and 6000 **3921**

2000 and 754 **1246**   3105 and 7000. **3895**

Subtraction • Difference between a multiple of 1000 and a four-digit number

---

Pupil Sheet 9

Name:

1

1 × 32 = **32**	3 × 32 = **64** + **32** = **96**
2 × 32 = **64**	6 × 32 = **128** + **64** = **192**
4 × 32 = **128**	7 × 32 = **64** + **32** + **128** = **224**
8 × 32 = **256**	11 × 32 = **256** + **64** + **32** = **352**

1 × 26 = **26**	5 × 26 = **104** + **26** = **130**
2 × 26 = **52**	9 × 26 = **208** + **26** = **234**
4 × 26 = **104**	7 × 26 = **104** + **52** + **26** = **182**
8 × 26 = **208**	13 × 26 = **208** + **104** + **26** = **338**

2

1 × 17	1 × 10 = 10 +	1 × 7 = 7	= **17**
2 × 17	2 × 10 = 20 +	2 × 7 = 14	= **34**
3 × 17	3 × 10 = 30 +	3 × 7 = 21	= **51**
4 × 17	4 × 10 = 40 +	4 × 7 = 28	= **68**
5 × 17	5 × 10 = 50 +	5 × 7 = 35	= **85**
6 × 17	6 × 10 = 60 +	6 × 7 = 42	= **102**
7 × 17	7 × 10 = 70 +	7 × 7 = 49	= **119**
8 × 17	8 × 10 = 80 +	8 × 7 = 56	= **136**
9 × 17	9 × 10 = 90 +	9 × 7 = 63	= **153**
10 × 17	10 × 10 = 100 +	10 × 7 = 70	= **170**

Multiplication • Generating tables

Pupil Sheet 10

Name:

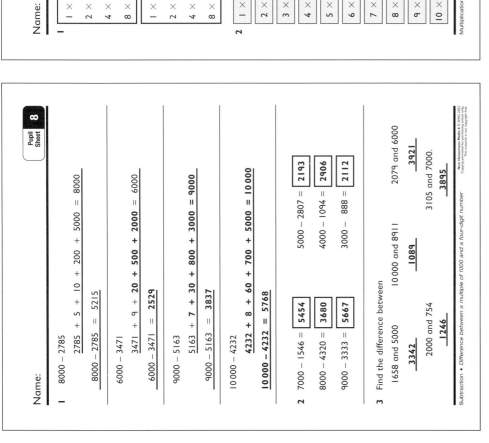

5 × 26 → 10 × 13 = **130**   5 × 18 → 10 × 9 = **90**   35 × 16 → 70 × 8 = **560**

15 × 16 → 30 × 8 = **240**   45 × 14 → 90 × 7 = **630**   14 × 17 → 7 × 34 = **238**

19 × 12 → 38 × 6 = **228**   14 × 23 → 7 × 46 = **322**   50 × 18 → 100 × 9 = **900**

16 × 32 → 8 × 64 = **512**   29 × 12 → 58 × 6 = **348**   26 × 14 → 52 × 7 = **364**

Multiplication • By doubling and halving

Pupil Sheet 11

Name:

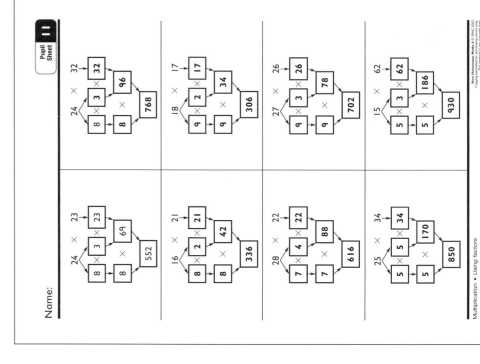

Multiplication • Using factors

Pupil Sheets

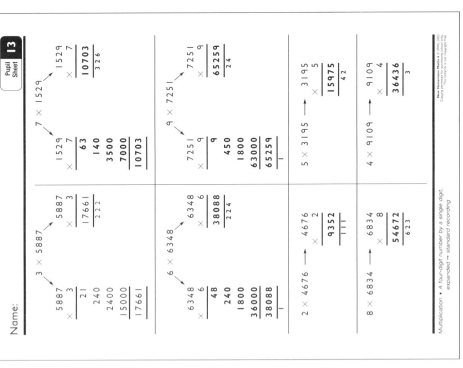

Pupil Sheet 13

Name:

3 × 5887:
```
  5887        5887
×    3      ×    3
    21       17661
   240        2 2 2
  2400
 15000
 17661
```

7 × 1529:
```
  1529        1529
×    7      ×    7
    63       10703
   140        3 2 6
  3500
  7000
 10703
```

6 × 6348:
```
  6348        6348
×    6      ×    6
    48       38088
   240        2 2 4
  1800
 36000
 38088
```

9 × 7251:
```
  7251        7251
×    9      ×    9
     9       65259
   450         2 4
  1800
 63000
 65259
```

2 × 4676:
```
 4676
×   2
 9352
 1 1 1
```

5 × 3195:
```
  3195
×    5
 15975
   4 2
```

8 × 6834:
```
  6834
×    8
 54672
 6 2 3
```

4 × 9109:
```
  9109
×    4
 36436
     3
```

Multiplication • A four-digit number by a single digit, expanded → standard recording

New Heinemann Maths 6 © SPMG 2002. Copying permitted for purchasing school only. This material is not copyright free.

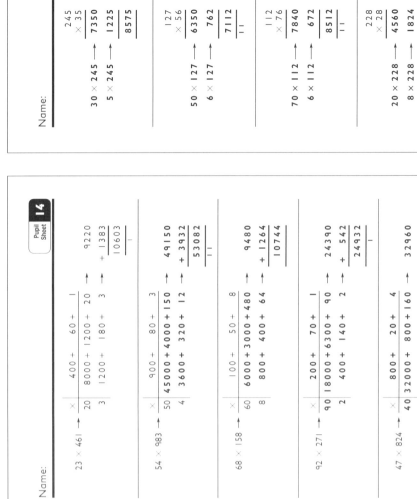

Pupil Sheet 15

Name:

```
   338
 ×  29
  6760   ← 20 × 338
  3042   ←  9 × 338
  9802
     1
```

```
   245
 ×  35
  7350   ← 30 × 245
  1225   ←  5 × 245
  8575
   1 1
```

```
   155
 ×  37
  4650   ← 30 × 155
  1085   ←  7 × 155
  5735
     1
```

```
   127
 ×  56
  6350   ← 50 × 127
   762   ←  6 × 127
  7112
   1 1
```

```
   146
 ×  62
  8760   ← 60 × 146
   292   ←  2 × 146
  9052
   1 1
```

```
   112
 ×  76
  7840   ← 70 × 112
   672   ←  6 × 112
  8512
   1 1
```

```
   416
 ×  24
  8320   ← 20 × 416
  1664   ←  4 × 416
  9984
     1
```

```
   228
 ×  28
  4560   ← 20 × 228
  1824   ←  8 × 228
  6384
     1
```

Multiplication • A three-digit number by a two-digit number, standard recording

New Heinemann Maths 6 © SPMG 2002. Copying permitted for purchasing school only. This material is not copyright free.

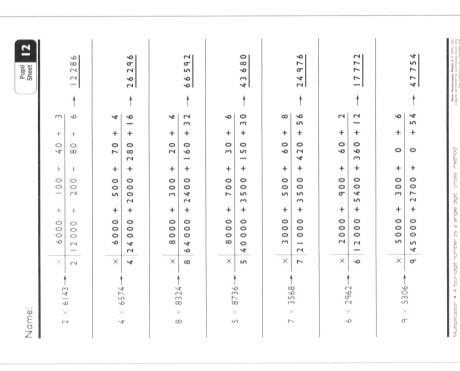

Pupil Sheet 12

Name:

```
2 × 6143 →    6000 +  100 +  40 +  3
           ×  2
          12000 +  200 +  80 +  6   → 12286
```

```
4 × 6574 →    6000 +  500 +  70 +   4
           ×  4
          24000 + 2000 + 280 +  16  → 26296
```

```
8 × 8324 →    8000 +  300 +  20 +   4
           ×  8
          64000 + 2400 + 160 +  32  → 66592
```

```
5 × 8736 →    8000 +  700 +  30 +   6
           ×  5
          40000 + 3500 + 150 +  30  → 43680
```

```
7 × 3568 →    3000 +  500 +  60 +   8
           ×  7
          21000 + 3500 + 420 +  56  → 24976
```

```
6 × 2962 →    2000 +  900 +  60 +   2
           ×  6
          12000 + 5400 + 360 +  12  → 17772
```

```
9 × 5306 →    5000 +  300 +   0 +   6
           ×  9
          45000 + 2700 +   0 +  54  → 47754
```

Multiplication • A four-digit number by a single digit: 'cross' method

New Heinemann Maths 6 © SPMG 2002. Copying permitted for purchasing school only. This material is not copyright free.

Pupil Sheet 14

Name:

```
23 × 461 → ×    400 + 60 +  1
         20   8000 + 1200 + 20  → 9220
          3   1200 +  180 +  3  → 1383
                                  10603
                                   1 1
```

```
54 × 983 → ×    900 +  80 +   3
         50  45000 + 4000 + 150 → 49150
          4   3600 +  320 +  12 → 3932
                                  53082
                                   1 1
```

```
68 × 158 → ×    100 +  50 +   8
         60   6000 + 3000 + 480 → 9480
          8    800 +  400 +  64 → 1264
                                  10744
                                   1 1
```

```
92 × 271 → ×    200 +  70 +   1
         90  18000 + 6300 +  90 → 24390
          2    400 +  140 +   2 → 542
                                  24932
                                     1
```

```
47 × 824 → ×    800 +  20 +   4
         40  32000 +  800 + 160 → 32960
          7   5600 +  140 +  28 → 5768
                                  38728
                                   1 1
```

Multiplication • A three-digit number by a two-digit number: 'cross' method

New Heinemann Maths 6 © SPMG 2002. Copying permitted for purchasing school only. This material is not copyright free.

Pupil Sheets

Pupil Sheet 17

Name: _____

1

half of 436	half of 400 = 200 / half of 36 = 18	→ 218
half of 672	half of 600 = 300 / half of 72 = 36	→ 336
half of 528	half of 500 = 250 / half of 28 = 14	→ 264
half of 784	half of 700 = 350 / half of 84 = 42	→ 392
half of 1160	half of 1100 = 550 / half of 60 = 30	→ 580
half of 1320	half of 1300 = 650 / half of 20 = 10	→ 660
half of 1580	half of 1500 = 750 / half of 80 = 40	→ 790

2 Find half of

904 → 452   732 → 366   1098 → 549
1120 → 560   1940 → 970   1760 → 880
4600 → 2300   8200 → 4100   3400 → 1700
5800 → 2900   7400 → 3700   9600 → 4800

Division • Halving

New Heinemann Maths 6 © SPMG 2002
Copying permitted for purchasing school only.
This material is not copyright free.

Pupil Sheet 20

Name: _____

1 536 nails are shared equally among 4 packs.
How many are in each pack?

```
  1 3 4
4 ) 5 3⁶6
```
→ 134 nails

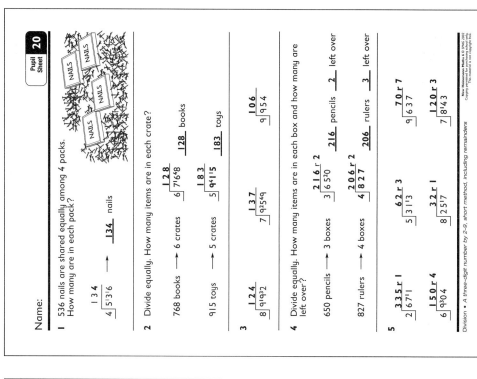

2 Divide equally. How many items are in each crate?

768 books → 6 crates
```
  1 2 8
6 ) 7 1⁶4⁴8
```
128 books

915 toys → 5 crates
```
  1 8 3
5 ) 9 ⁴4¹5
```
183 toys

3
```
  1 2 4
8 ) 9 ⁱ9³2
```
```
  1 3 7
7 ) 9 ²5⁴9
```
```
  1 0 6
9 ) 9 5 4
```

4 Divide equally. How many items are in each box and how many are left over?

650 pencils → 3 boxes
```
  2 1 6 r 2
3 ) 6 ³5⁰0
```
216 pencils   2 left over

827 rulers → 4 boxes
```
  2 0 6 r 2
4 ) 8 2⁷7
```
206 rulers   3 left over

5
```
  3 3 5 r 1
2 ) 6 ⁷1
```
```
  6 2 r 3
5 ) 3 ¹3
```
```
  7 0 r 7
9 ) 6 3 7
```
```
  1 5 0 r 4
6 ) 9 0 4
```
```
  3 2 r 1
8 ) 2 ⁵7
```
```
  1 2 0 r 3
7 ) 8 1 4 3
```

Division • A three-digit number by 2–9, short method, including remainders

New Heinemann Maths 6 © SPMG 2002
Copying permitted for purchasing school only.
This material is not copyright free.

Pupil Sheet 16

Name: _____

Mark each child's work.
Tick (✓) correct answers. Cross (X) wrong answers.

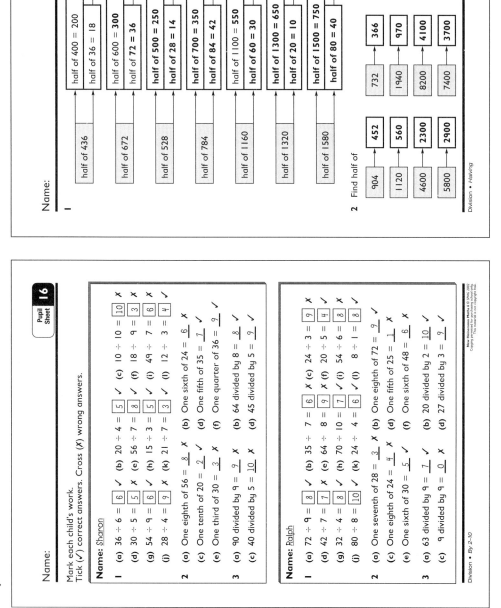

Name: Sharon

1 (a) 36 ÷ 6 = [6] ✓ (b) 20 ÷ 4 = [5] ✓ (c) 10 ÷ 10 = [10] X
(d) 30 ÷ 5 = [5] X (e) 56 ÷ 7 = [8] ✓ (f) 18 ÷ 9 = [3] ✓
(g) 54 ÷ 9 = [6] ✓ (h) 15 ÷ 3 = [5] ✓ (i) 49 ÷ 7 = [6] X
(j) 28 ÷ 4 = [9] X (k) 21 ÷ 7 = [3] ✓ (l) 12 ÷ 3 = [4] ✓

2 (a) One eighth of 56 = 8 X (b) One sixth of 24 = 6 X
(c) One tenth of 20 = 2 ✓ (d) One fifth of 35 = 7 ✓
(e) One third of 30 = 3 X (f) One quarter of 36 = 9 ✓

3 (a) 90 divided by 9 = 9 X (b) 64 divided by 8 = 8 ✓
(c) 40 divided by 5 = 10 X (d) 45 divided by 5 = 9 ✓

Name: Ralph

1 (a) 72 ÷ 9 = [8] ✓ (b) 35 ÷ 7 = [6] X (c) 24 ÷ 3 = [9] ✓
(d) 42 ÷ 7 = [7] X (e) 64 ÷ 8 = [9] ✓ (f) 20 ÷ 5 = [4] X
(g) 32 ÷ 4 = [8] ✓ (h) 70 ÷ 10 = [7] ✓ (i) 54 ÷ 6 = [8] X
(j) 80 ÷ 8 = [10] X (k) 24 ÷ 4 = [6] ✓ (l) 8 ÷ 1 = [8] ✓

2 (a) One seventh of 28 = 3 X (b) One eighth of 72 = 9 ✓
(c) One eighth of 24 = 4 X (d) One fifth of 25 = 1 ✓
(e) One sixth of 30 = 5 ✓ (f) One sixth of 48 = 6 ✓

3 (a) 63 divided by 9 = 7 ✓ (b) 20 divided by 2 = 10 ✓
(c) 9 divided by 9 = 0 X (d) 27 divided by 3 = 9 ✓

Division • By 2–10

New Heinemann Maths 6 © SPMG 2002
Copying permitted for purchasing school only.
This material is not copyright free.

Pupil Sheet 18

Name: _____

```
7 ) 798
   -700    7 × 100
     98
    -70    7 × 10
     28
    -28    7 × 4
      0
798 ÷ 7 = 114
```

```
6 ) 864
   -600    6 × 100
    264
   -240    6 × 40
     24
    -24    6 × 4
      0
864 ÷ 6 = 144
```

```
4 ) 936
   -800    4 × 200
    136
   -120    4 × 30
     16
    -16    4 × 4
      0
936 ÷ 4 = 234
```

```
8 ) 899
   -800    8 × 100
     99
    -80    8 × 10
     19
    -16    8 × 2
      3
899 ÷ 8 = 112 r 3
```

```
5 ) 783
   -500    5 × 100
    283
   -250    5 × 50
     33
    -30    5 × 6
      3
783 ÷ 5 = 156 r 3
```

```
3 ) 685
   -600    3 × 200
     85
    -60    3 × 20
     25
    -24    3 × 8
      1
685 ÷ 3 = 228 r 1
```

Division • By 2–9, three-digit quotients, including remainders

New Heinemann Maths 6 © SPMG 2002
Copying permitted for purchasing school only.
This material is not copyright free.

Pupil Sheets

Pupil Sheet 21

Name:

1 511 apples are shared equally among 52 boxes. **About** how many apples are in each box?

511 ÷ 52 is **approximately** 500 ÷ 50 = __10__ apples.

2 How many classes of 28 can be made with 587 children?

587 ÷ 28 is **approximately** 600 ÷ 30 = __20__ classes.

3 Find an approximate answer for each division.

822 ÷ 39 → 800 ÷ 40 = __20__

393 ÷ 17 → 400 ÷ 20 = __20__

609 ÷ 22 → 600 ÷ 20 = __30__

889 ÷ 33 → 900 ÷ 30 = __30__

978 ÷ 21 → 1000 ÷ 20 = __50__

521 ÷ 18 → 500 ÷ 20 = __25__

993 ÷ 51 → 1000 ÷ 50 = __20__

428 ÷ 42 → 400 ÷ 40 = __10__

786 ÷ 23 → 800 ÷ 20 = __40__

372 ÷ 48 → 400 ÷ 50 = __8__

288 ÷ 27 → 300 ÷ 30 = __10__

Division • A three-digit number by a two-digit number: approximate quotients

---

Pupil Sheet 22

Name:

```
26|546
  -260    26 × 10
   286
  -260    26 × 10
    26
   -26    26 × 1
     0
546 ÷ 26 = 21
```

```
18|396
  -180    18 × 10
   216
  -180    18 × 10
    36
   -36    18 × 2
     0
396 ÷ 18 = 22
```

```
22|682
  -220    22 × 10
   462
  -220    22 × 10
   242
  -220    22 × 10
    22
   -22    22 × 1
     0
682 ÷ 22 = 31
```

```
24|840
  -240    24 × 10
   600
  -240    24 × 10
   360
  -240    24 × 10
   120
  -120    24 × 5
     0
840 ÷ 24 = 35
```

```
33|736
  -330    33 × 10
   406
  -330    33 × 10
    76
   -66    33 × 2
    10
736 ÷ 33 = 22 r 10
```

```
28|777
  -280    28 × 10
   497
  -280    28 × 10
   217
  -140    28 × 5
    77
   -56    28 × 2
    21
777 ÷ 28 = 27 r 21
```

Division • A three-digit number by a two-digit number, including remainders

---

Pupil Sheet 23

Name:

```
21|462
  420    21 × 20
   42
   42    21 × 2
    0
462 ÷ 21 = 22
```

```
42|966
  840    42 × 20
  126
  126    42 × 3
    0
966 ÷ 42 = 23
```

```
31|961
  930    31 × 30
   31
   31    31 × 1
    0
961 ÷ 31 = 31
```

```
26|910
  780    26 × 30
  130
  130    26 × 5
    0
910 ÷ 26 = 35
```

```
17|415
  340    17 × 20
   75
   68    17 × 4
    7
415 ÷ 17 = 24 r 7
```

```
35|888
  700    35 × 20
  188
  175    35 × 5
   13
888 ÷ 35 = 25 r 13
```

```
24|776
  720    24 × 30
   56
   48    24 × 2
    8
776 ÷ 24 = 32 r 8
```

```
22|949
  880    22 × 40
   69
   66    22 × 3
    3
949 ÷ 22 = 43 r 3
```

Division • A three-digit number by a two-digit number, including remainders: shorter method

---

Pupil Sheet 25

Name:

1 (a) Complete this number sequence.

Add 4   0   4   8   12   16

__20__   __24__   __28__   __32__   __36__

(b) The pattern of the **units digits** is:

0 4 8 2 6 0 4 ...

Show this pattern on the circle by joining the numbered dots in order.

(c) Repeat for a different starting number.

Add 4 ____ : ____ : ____ : ...

2 Investigate circle patterns for other number sequences.

Add 2 ____ : ____ : ...

Add 7 ____ : ____ : ...

Number properties • Last digit patterns

Pupil Sheets

Name:

1 (a) Cross out 1 and 49 in the number square.
(b) Cross out all the multiples
  • of 2, except 2
  • of 3, except 3
  • of 5, except 5.

2 The numbers **not** crossed out are **prime numbers**.
List these prime numbers.

2 , 3 , 5 , 7 , 11 ,
13 , 17 , 19 , 23 , 29 , 31 , 37 , 41 , 43 , 47

Number properties • Prime numbers

Name:

1 Complete.

$1\frac{5}{8} = \frac{13}{8}$

$4\frac{3}{4} = \frac{19}{4}$

$2\frac{5}{6} = \frac{17}{6}$

$1\frac{2}{5} = \frac{7}{5}$

$3\frac{1}{3} = \frac{10}{3}$

$1\frac{6}{10} = \frac{16}{10}$

2 Change to improper fractions.

$2\frac{2}{3} \to \frac{8}{3}$ $3\frac{1}{4} \to \frac{13}{4}$ $5\frac{3}{7} \to \frac{38}{7}$ $4\frac{3}{10} \to \frac{43}{10}$

$7\frac{4}{9} \to \frac{67}{9}$ $6\frac{4}{6} \to \frac{40}{6}$ $9\frac{1}{2} \to \frac{19}{2}$ $2\frac{7}{12} \to \frac{31}{12}$

3 Change to mixed numbers.

$\frac{9}{5} \to 1\frac{4}{5}$ $\frac{23}{8} \to 2\frac{7}{8}$ $\frac{34}{7} \to 4\frac{6}{7}$ $\frac{49}{12} \to 4\frac{1}{12}$

$\frac{69}{20} \to 3\frac{9}{20}$ $\frac{51}{6} \to 8\frac{3}{6}$ $\frac{62}{8} \to 7\frac{6}{8}$ $\frac{90}{10} \to 9$

Fractions • Mixed numbers and improper fractions

Name:

Shade the second shape in each pair to match the first one.
Complete each equal fractions story.

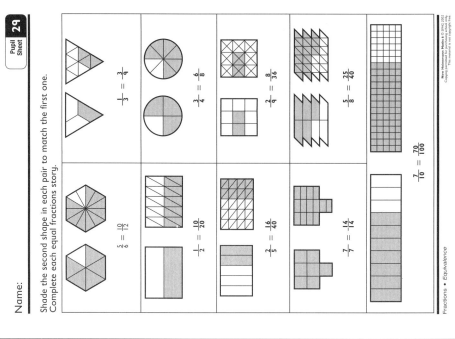

$\frac{1}{3} = \frac{3}{9}$

$\frac{3}{4} = \frac{6}{8}$

$\frac{2}{9} = \frac{8}{36}$

$\frac{5}{8} = \frac{25}{40}$

$\frac{5}{6} = \frac{10}{12}$

$\frac{1}{2} = \frac{10}{20}$

$\frac{2}{5} = \frac{16}{40}$

$\frac{7}{7} = \frac{14}{14}$

$\frac{7}{10} = \frac{70}{100}$

Fractions • Equivalence

Name:

Shade the second shape in each pair to match the first one.
Complete each equal fractions story.

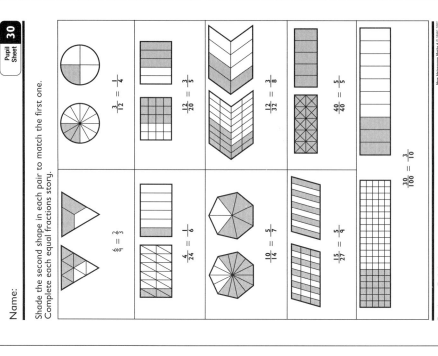

$\frac{3}{12} = \frac{1}{4}$

$\frac{12}{20} = \frac{3}{5}$

$\frac{12}{32} = \frac{3}{8}$

$\frac{40}{40} = \frac{5}{5}$

$\frac{6}{9} = \frac{2}{3}$

$\frac{4}{24} = \frac{1}{6}$

$\frac{10}{14} = \frac{5}{7}$

$\frac{15}{27} = \frac{5}{9}$

$\frac{30}{100} = \frac{3}{10}$

Fractions • Equivalence

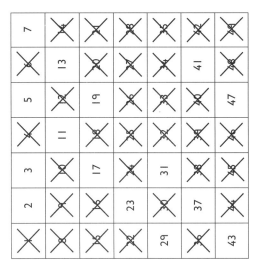

Pupil Sheets

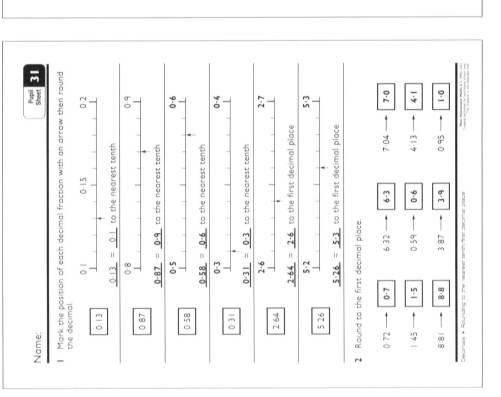

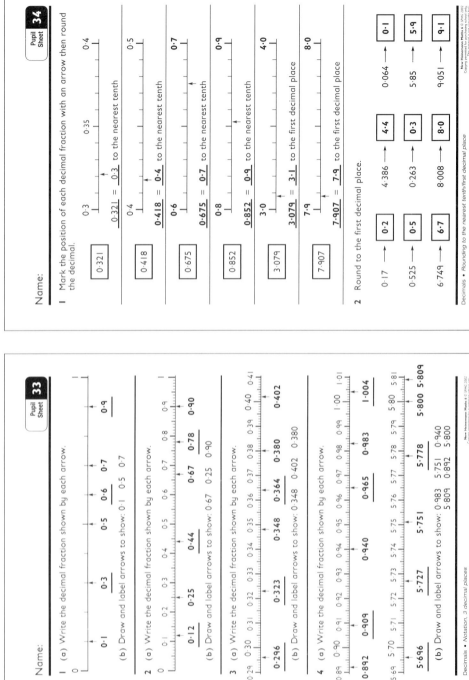

Pupil Sheet 31

Name:

1 Mark the position of each decimal fraction with an arrow then round the decimal.

| 0.13 | 0.1   0.15   0.2 |
0.13 = 0.1 to the nearest tenth

| 0.87 | 0.8   0.9 |
0.87 = 0.9 to the nearest tenth

| 0.58 | 0.5   0.6 |
0.58 = 0.6 to the nearest tenth

| 0.31 | 0.3   0.4 |
0.31 = 0.3 to the nearest tenth

| 2.64 | 2.6   2.7 |
2.64 = 2.6 to the first decimal place

| 5.26 | 5.2   5.3 |
5.26 = 5.3 to the first decimal place

2 Round to the first decimal place.

0.72 →	0.7		6.32 →	6.3		7.04 →	7.0
1.45 →	1.5		0.59 →	0.6		4.13 →	4.1
8.81 →	8.8		3.87 →	3.9		0.95 →	1.0

Decimals • Rounding to the nearest tenth/first decimal place

New Heinemann Maths 6, © SPMG 2002
Copying permitted for purchasing school only.
This material is not copyright free.

Pupil Sheet 33

Name:

1 (a) Write the decimal fraction shown by each arrow.

0.3   0.5   0.6   0.7   0.9

(b) Draw and label arrows to show: 0.1   0.5   0.7

2 (a) Write the decimal fraction shown by each arrow.

0   0.1   0.2   0.3   0.4   0.5   0.6   0.7   0.8   0.9

0.12   0.25   0.44   0.67   0.78   0.90

(b) Draw and label arrows to show: 0.67   0.25   0.90

3 (a) Write the decimal fraction shown by each arrow.

0.29   0.30   0.31   0.32   0.33   0.34   0.35   0.36   0.37   0.38   0.39   0.40   0.41

0.296   0.323   0.348   0.364   0.380   0.402

(b) Draw and label arrows to show: 0.348   0.402   0.380

4 (a) Write the decimal fraction shown by each arrow.

0.89   0.90   0.91   0.92   0.93   0.94   0.95   0.96   0.97   0.98   0.99   1.00   1.01

0.892   0.909   0.940   0.965   0.983   1.004

5.69   5.70   5.71   5.72   5.73   5.74   5.75   5.76   5.77   5.78   5.79   5.80   5.81

5.696   5.727   5.751   5.778   5.800   5.809

(b) Draw and label arrows to show: 0.983   5.751   0.940
5.809   0.892   5.800

Decimals • Notation, 3 decimal places

New Heinemann Maths 6, © SPMG 2002
Copying permitted for purchasing school only.
This material is not copyright free.

Pupil Sheet 34

Name:

1 Mark the position of each decimal fraction with an arrow then round the decimal.

| 0.321 | 0.3   0.35   0.4 |
0.321 = 0.3 to the nearest tenth

| 0.418 | 0.4   0.5 |
0.418 = 0.4 to the nearest tenth

| 0.675 | 0.6   0.7 |
0.675 = 0.7 to the nearest tenth

| 0.852 | 0.8   0.9 |
0.852 = 0.9 to the nearest tenth

| 3.079 | 3.0   4.0 |
3.079 = 3.1 to the first decimal place

| 7.907 | 7.9   8.0 |
7.907 = 7.9 to the first decimal place

2 Round to the first decimal place.

0.17 →	0.2		4.386 →	4.4		0.064 →	0.1
0.525 →	0.5		0.263 →	0.3		5.85 →	5.9
6.749 →	6.7		8.008 →	8.0		9.051 →	9.1

Decimals • Rounding to the nearest tenth/first decimal place

New Heinemann Maths 6, © SPMG 2002
Copying permitted for purchasing school only.
This material is not copyright free.

Pupil Sheets

Pupil Sheet 35

Name:

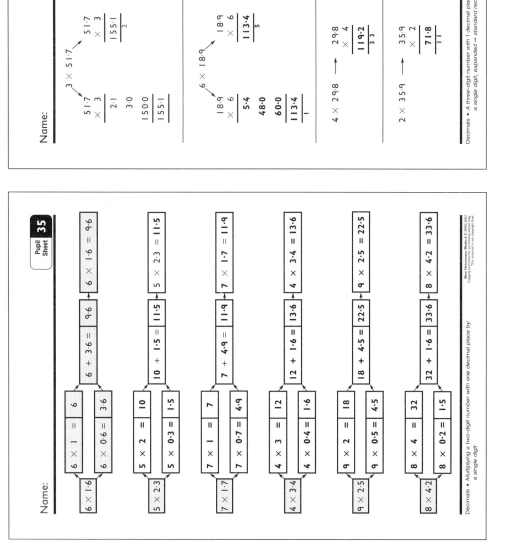

6 × 1·6 | 6 × 1 = 6 | 6 × 0·6 = 3·6 | 6 + 3·6 = 9·6 | 6 × 1·6 = 9·6

5 × 2·3 | 5 × 2 = 10 | 5 × 0·3 = 1·5 | 10 + 1·5 = 11·5 | 5 × 2·3 = 11·5

7 × 1·7 | 7 × 1 = 7 | 7 × 0·7 = 4·9 | 7 + 4·9 = 11·9 | 7 × 1·7 = 11·9

4 × 3·4 | 4 × 3 = 12 | 4 × 0·4 = 1·6 | 12 + 1·6 = 13·6 | 4 × 3·4 = 13·6

9 × 2·5 | 9 × 2 = 18 | 9 × 0·5 = 4·5 | 18 + 4·5 = 22·5 | 9 × 2·5 = 22·5

8 × 4·2 | 8 × 4 = 32 | 8 × 0·2 = 1·6 | 32 + 1·6 = 33·6 | 8 × 4·2 = 33·6

Decimals ● Multiplying a two-digit number with one decimal place by a single digit

---

Pupil Sheet 36

Name:

3 × 51·7
```
  51·7        51·7
×    3   →  ×    3
   2·1      155·1
   3·0         2
 150·0
 155·1
```

5 × 14·3
```
  14·3        14·3
×    5   →  ×    5
   1·5       71·5
  20·0          1
  50·0
  71·5
```

8 × 26·1
```
  26·1        26·1
×    8   →  ×    8
   0·8      208·8
  48·0          1
 160·0
 208·8
```

6 × 18·9
```
  18·9        18·9
×    6   →  ×    6
   5·4      113·4
  48·0          5
  60·0
 113·4
```

4 × 29·8
```
  29·8       29·8
×    4   →  ×   4
 119·2      119·2
               3 3
```

7 × 36·4
```
  36·4       36·4
×    7   →  ×   7
            254·8
              4 2
```

9 × 40·6
```
  40·6       40·6
×    9   →  ×   9
            365·4
                5
```

2 × 35·9
```
  35·9       35·9
×    2   →  ×   2
            71·8
             1 1
```

Decimals ● A three-digit number with 1 decimal place multiplied by a single digit, expanded → standard recording

---

Pupil Sheet 37

Name:

4 × 1·63
```
  1·63        1·63
×    4   →  ×    4
  0·12        6·52
  2·40          2 1
  4·00
  6·52
```

7 × 2·74
```
  2·74        2·74
×    7   →  ×    7
  0·28       19·18
  4·90          5 2
 14·00
 19·18
```

9 × 3·52
```
  3·52        3·52
×    9   →  ×    9
  0·18       31·68
  4·50          4 1
 27·00
 31·68
```

6 × 4·43
```
  4·43        4·43
×    6   →  ×    6
  0·18       26·58
  2·40          2 1
 24·00
 26·58
```

2 × 6·37
```
  6·37       6·37
×    2   →  ×   2
            12·74
               1
```

5 × 3·95
```
  3·95       3·95
×    5   →  ×   5
            19·75
              4 2
```

3 × 2·08
```
  2·08       2·08
×    3   →  ×   3
             6·24
                2
```

8 × 1·46
```
  1·46       1·46
×    8   →  ×   8
            11·68
              3 4
```

Decimals ● A three-digit number with 2 decimal places multiplied by a single digit, expanded → standard recording

---

Pupil Sheet 38

Name:

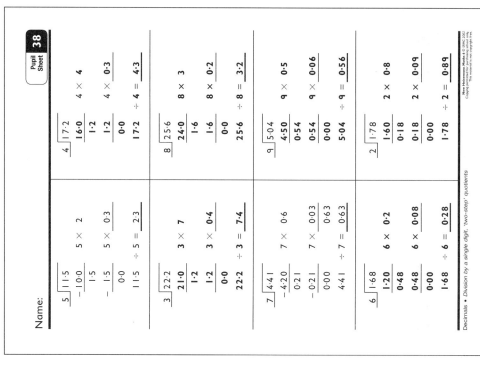

```
     2·3
 5 ) 11·5
  - 10·0     5 × 2
     1·5
   - 1·5     5 × 0·3
     0·0
  11·5 ÷ 5 = 2·3
```

```
     4·3
 4 ) 17·2
    16·0     4 × 4
     1·2
     1·2     4 × 0·3
     0·0
  17·2 ÷ 4 = 4·3
```

```
     7·4
 3 ) 22·2
    21·0     3 × 7
     1·2
     1·2     3 × 0·4
     0·0
  22·2 ÷ 3 = 7·4
```

```
     3·2
 8 ) 25·6
    24·0     8 × 3
     1·6
     1·6     8 × 0·2
     0·0
  25·6 ÷ 8 = 3·2
```

```
     0·63
 7 ) 4·41
   - 4·20    7 × 0·6
     0·21
   - 0·21    7 × 0·03
     0·00
  4·41 ÷ 7 = 0·63
```

```
     0·56
 9 ) 5·04
     4·50    9 × 0·5
     0·54
     0·54    9 × 0·06
     0·00
  5·04 ÷ 9 = 0·56
```

```
     0·28
 6 ) 1·68
     1·20    6 × 0·2
     0·48
     0·48    6 × 0·08
     0·00
  1·68 ÷ 6 = 0·28
```

```
     0·89
 2 ) 1·78
     1·60    2 × 0·8
     0·18
     0·18    2 × 0·09
     0·00
  1·78 ÷ 2 = 0·89
```

Decimals ● Division by a single digit, 'two-step' quotients

Pupil Sheets

Pupil Sheet 40

Name:

3 | 9 4 2
- 90·0   3 × 30
   4 2
- 3·0   3 × 1
   1 2
- 1·2   3 × 0·4
   0·0
9 4 2 ÷ 3 = 31·4

2 | 3 3·4
- 30·0   2 × 15
   3·4
- 3·0   2 × 1·5
   0·4
- 0·4   2 × 0·2
   0·0
33·4 ÷ 2 = 16·7

6 | 8 3 4
- 60·0   6 × 10
  23·4
- 18·0   6 × 3·0
   5·4
- 5·4   6 × 0·9
   0·0
83·4 ÷ 6 = 13·9

4 | 8·68
- 8·00   4 × 2
   0·68
- 0·40   4 × 0·1
   0·28
- 0·28   4 × 0·07
   0·00
8·68 ÷ 4 = 2·17

5 | 6 25
- 5·00   5 × 1
   1·25
- 1·00   5 × 0·2
   0·25
- 0·25   5 × 0·05
   0·00
6·25 ÷ 5 = 1·25

7 | 9·87
- 7·00   7 × 1
   2·87
- 2·80   7 × 0·4
   0·07
- 0·07   7 × 0·01
   0·00
9·87 ÷ 7 = 1·41

Decimals • Division by a single digit, three-step quotients

Pupil Sheet 42

Name:

Colour.

65%

36%

80%

60%

50%

90%

25%

75%

40%

Percentages • Of a shape

Pupil Sheet 43

Name:

Write each weight
• to the nearest 50g

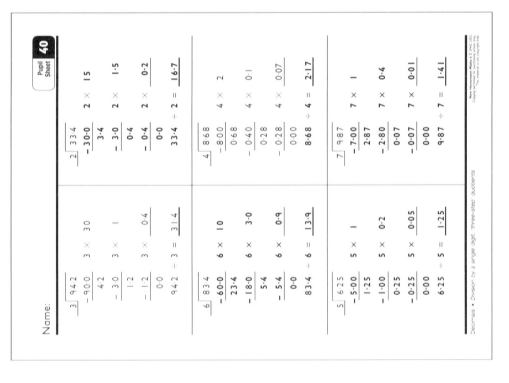

250 g   750 g   1200 g

• to the nearest 20g

360 g   560 g   1020 g

• to the nearest 10g

150 g   370 g   610 g

Weight • Reading scales to the nearest mark, 50 g, 20 g, 10 g divisions

Pupil Sheet 44

Name:

1   Write each weight • in kilograms, to the nearest $\frac{1}{10}$ kg
                  • in grams, to the nearest 100 g.

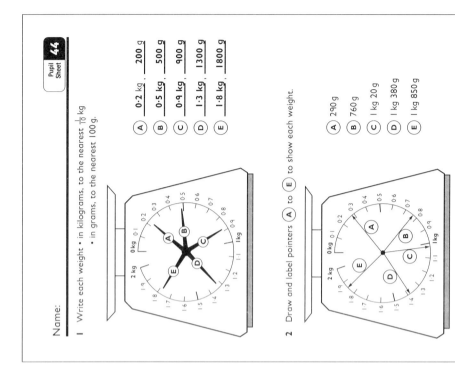

A   0·2 kg   200 g
B   0·5 kg   500 g
C   0·9 kg   900 g
D   1·3 kg   1300 g
E   1·8 kg   1800 g

2   Draw and label pointers A to E to show each weight.

A   290 g
B   760 g
C   1 kg 20 g
D   1 kg 380 g
E   1 kg 850 g

Weight • Reading scales to the nearest mark, tenths of 1 kg divisions

Pupil Sheets

Name:

Find the area of each shape.

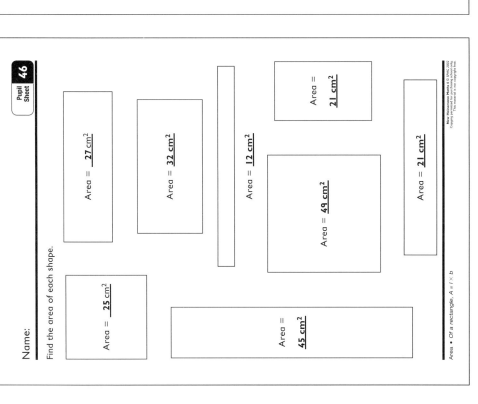

Area = __25 cm²__

Area = __27 cm²__

Area = __32 cm²__

Area = __12 cm²__

Area = __49 cm²__

Area = __21 cm²__

Area = __45 cm²__

Area = __21 cm²__

Name:

1 Tick (✓) shapes that have an area of **4** square units.

2 Complete each shape so that it has an area of **6** square units.
   Colour your shapes.

Name:

Sketch the reflection of each shape.

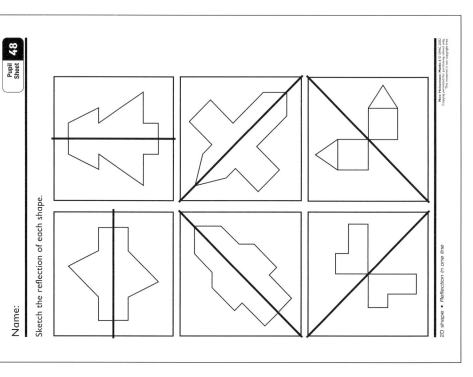

Pupil Sheets

Pupil Sheet 54

Name:

1 Name each shape.

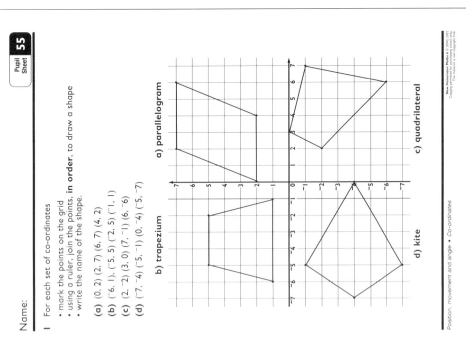

parallelogram

rectangle

kite

rhombus

square

isosceles trapezium

quadrilateral

2 Use a ruler. Draw the diagonals of each shape.

3 List the names of shapes whose diagonals
• intersect at right angles

kite, square, rhombus

• bisect each other.

parallelogram, rectangle, isosceles trapezium, quadrilateral

4 In which of the shapes is
• each diagonal a line of symmetry   **square, rhombus**

• only one diagonal a line of symmetry?   **kite**

New Heinemann Maths 6 © JMPG 2002
Copying permitted for purchasing school only.
This material is not copyright free

2D Shape • Diagonal properties

Pupil Sheet 55

Name:

1 For each set of co-ordinates
• mark the points on the grid
• using a ruler, join the points, **in order**, to draw a shape
• write the name of the shape.

(a) (0, 2) (2, 7) (6, 7) (4, 2)
(b) (⁻6, 1), (⁻5, 5) (2, 5) (⁻1, 1)
(c) (2, ⁻2) (3, 0) (7, ⁻1) (6, ⁻6)
(d) (⁻7, ⁻4) (⁻5, ⁻1) (0, ⁻4) (⁻5, ⁻7)

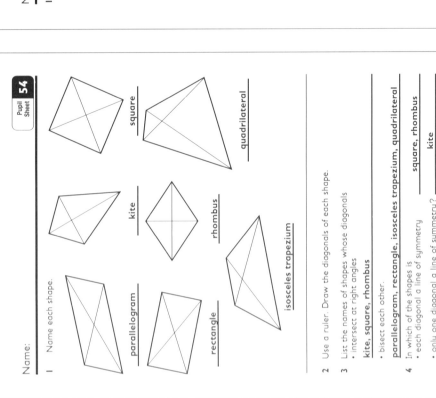

a) parallelogram

b) trapezium

c) quadrilateral

d) kite

New Heinemann Maths 6 © JMPG 2002
Copying permitted for purchasing school only.
This material is not copyright free

Position, movement and angle • Co-ordinates

Pupil Sheet 56

Name:

Sketch the new position of each shape after it has rotated
about the vertex at the centre, marked with a dot.

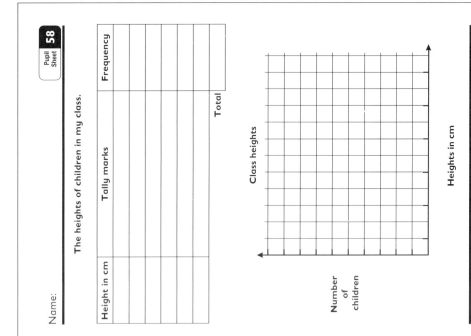

90° anti-clockwise

90° clockwise

180° clockwise

180° anti-clockwise

New Heinemann Maths 6 © JMPG 2002
Copying permitted for purchasing school only.
This material is not copyright free

Position, movement and angle • Rotation

Pupil Sheet 58

Name:

The heights of children in my class.

Height in cm	Tally marks	Frequency
		Total

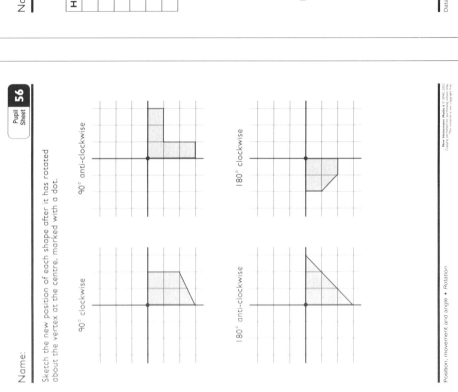

Class heights

Number
of
children

Heights in cm

New Heinemann Maths 6 © JMPG 2002
Copying permitted for purchasing school only.
This material is not copyright free

Data handling • Bar chart sheet (for use with Textbook page 121)

Assessment

Name:

What is the value of the 6 in each number?

4 826 159	six thousands
3 507 628	six hundreds
13 637 829	six hundred thousands

Write in order. Start with the smallest number.

5 764 125   3 856 214   3 867 124   5 768 412   3 858 421

3 856 214   3 858 421   3 867 124   5 764 125   5 768 412

Write in words.

68 004   **Sixty-eight thousand and four.**

1 150 030   **One million, one hundred and fifty thousand and thirty.**

463 907 + 10 000 = **473 907**   3 615 829 − 100 000 = **3 515 829**

164 000 + 30 000 = **194 000**   45 082 + 4 000 = **49 082**

5 868 425 − 500 000 = **5 368 425**   778 073 − 60 000 = **718 073**

Find the cost of

10 Yachts **£1370**   (Yacht £137)

100 Balloons **£4600**   (Hot-Air Balloon £46)

53 × 100 = **5300**   128 × 1000 = **128 000**

72 × **10** = 720   **1420** × 100 = 142 000

65 000 ÷ 1000 = **65**   4900 ÷ 100 = **49**

804 000 ÷ 1000 = 804   90 700 ÷ **10** = 9070

Name:

Estimate the distance, in metres, from the start of the line to

• each arrow

• each ball.

0 m   200 m   400 m   700 m   1000 m

60 m   100 m   160 m   200 m

Mark and label arrows below the line to show the approximate position of each number.

8000   2900

0 .... 2900 .... 8000 10000

Round to the nearest thousand.

9318 → **9000**

148 673 → **149 000**

Round to the nearest hundred.

3281 → **3300**

49 945 → **49 900**

Round to the nearest million.

5 200 000   **5 000 000**

6 505 000   **7 000 000**

8 099 000   **8 000 000**

29 690 014   **30 000 000**

Name:

60 + 90 + 60 + 50 + 40 = **300**

28 + 16 + 34 + 41 = **119**

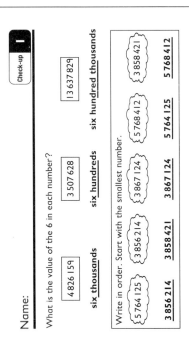

Find the total weight of the containers.

48kg   46kg   45kg   43kg   48kg   **230 kg**

536 + 348 = **884**   307 + 659 = **966**   155 + 826 = **981**

427 + 543 = **970**   619 + 274 = **893**   765 + 228 = **993**

329 + **437** = 766   **546** + 425 = 971

Find the total cost of

• the camera and the DVD player **£753**

• the DVD player and the camcorder **£839**

• the camcorder and the television **£958**

• the television and the camera. **£872**

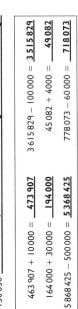
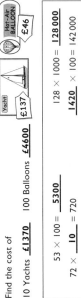

£366   £592   £473   £280

356 + **470** = 826   **283** + 280 = 563

570 + 820 = **1390**   360 + 921 = **1281**   608 + 767 = **1375**

734 + 652 = **1386**   417 + 682 = **1099**   545 + 644 = **1189**

882 + **811** = 1693   **536** + 530 = 1066

Name:

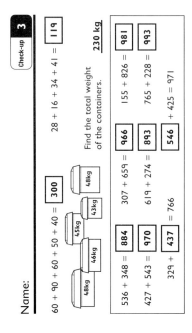

City £2400   Zest £4500   Alpha £3300   Trekk £5500   Mover £4900

What is the total cost of these cars?

• City and Alpha **£5700**   • Alpha and Trekk **£8800**

• Mover and City **£7300**   • Zest and Mover **£9400**

• Mover and Alpha **£8200**   • Trekk and Zest **£10 000**

7800 + 6100 = **13 900**   9600 + 6400 = **16 000**

5500 + 8700 = **14 200**   4800 + 6500 = **11 300**

2400 + **9200** = 11 600   9900 + **8800** = 18 700

Number of visitors to Castlebury Tower

April	May	June	July	August
1534	3354	3013	5626	4263

How many visitors altogether visited Castlebury Tower during

• April and May **4888**   • May and June **6367**

• June and July **8639**   • July and August? **9889**

8324 + 1538 = **9862**   7057 + 2635 = **9692**

5633 + 2259 = **7892**   1856 + 4136 = **5992**

Assessment

49

Check-up 6

Name:

6700 - 1500 = 5200
10300 - 4900 = 5400
4100 - 1500 = 2600
7500 - 1350 = 6150

8200 - 2700 = 5500
14000 - 9200 = 4800
5400 - 3600 = 1800
9400 - 3050 = 6350

BARNDALE BULLETIN
Sales Information

Month	Number of copies sold
January	8756
February	6103
March	9808
April	7524

Find the difference between the number of copies sold in
- January and February 2653
- February and March 3765
- April and January 1232
- January and March 1112
- April and February? 1421

March and April 2344

5340 - 2017 = 3323
4765 - 1329 = 3436

5000 - 3710 = 1290
6000 - 1555 = 4445
9000 - 4408 = 4592
8000 - 3642 = 4358

3000 - 981 = 2019
4000 - 693 = 3307
7000 - 2838 = 4162
10000 - 5961 = 4039

Subtraction: Pupil Sheet 6

Check-up 5

Name:

640 - 270 = 370
560 - 190 = 370
454 - 180 = 274
842 - 360 = 482

710 - 530 = 180
320 - 150 = 170
607 - 420 = 187
456 - 270 = 186

830 - 360 = 470
916 - 640 = 276

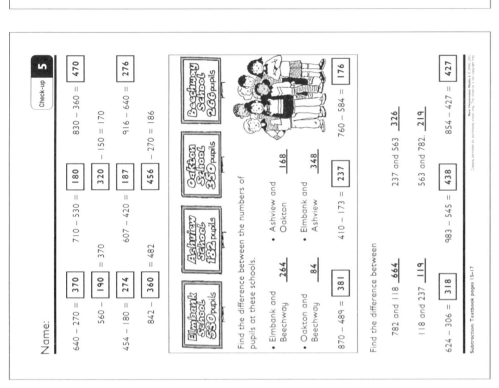

Beechway School 266 pupils
Oakton School 350 pupils
Ashview School 182 pupils
Elmbank School 530 pupils

Find the difference between the numbers of pupils at these schools.
- Elmbank and Beechway 264
- Ashview and Oakton 168
- Oakton and Beechway 84
- Elmbank and Ashview 348

870 - 489 = 381
410 - 173 = 237
760 - 584 = 176

Find the difference between
782 and 118 664
237 and 563 326
118 and 237 119
563 and 782 219

624 - 306 = 318
983 - 545 = 438
854 - 427 = 427

Subtraction: Textbook pages 15-17

Check-up 8

Name:

Match.

5 × 42   18 × 15   16 × 15   35 × 18   12 × 45
68 × 5   25 × 18   35 × 14

(210) (240) (270) (340) (450) (490) (540) (630)

16 × 17 = 272
47 × 12 = 564
36 × 50 = 1800
50 × 44 = 2200

31 × 14 = 434
16 × 27 = 432
25 × 32 = 800
28 × 25 = 700

18 × 23 = 414
33 × 14 = 462
52 × 25 = 1300
50 × 58 = 2900

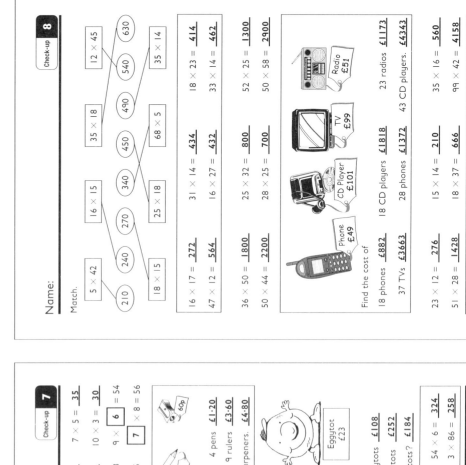

Radio £51   TV £99   CD Player £101   Phone £49

Find the cost of
18 phones £882
18 CD players £1818
28 phones £1372
37 TVs £3663

23 radios £1173
43 CD players £4343

23 × 12 = 276
51 × 28 = 1428

15 × 14 = 210
18 × 37 = 666

35 × 16 = 560
99 × 42 = 4158

Multiplication: Textbook pages 21-24

Check-up 7

Name:

3 × 8 = 24
9 × 7 = 63
6 × 4 = 24
8 × 8 = 64

7 × 6 = 42
8 × 4 = 32
7 × 7 = 49
9 × 3 = 27

4 × 9 = 36
6 × 6 = 36
4 × 7 = 28
5 × 5 = 25

7 × 5 = 35
10 × 3 = 30
6 × 9 = 54
7 × 8 = 56

Find the cost of
3 sharpeners £1·80
50 pencils £4·50
70 erasers £4·90

5 rulers £2
30 erasers £2·10
80 pencils £7·20

4 pens £1·20
9 rulers £3·60
8 sharpeners £4·80

Fairytot £28   Hairytot £36   Eggytot £23

What is the total cost of
4 Fairytots £112
7 Eggytots £161
5 Hairytots £180

3 Hairytots £108
9 Fairytots £252
8 Eggytots? £184

67 × 4 = 268
8 × 41 = 328

95 × 7 = 665
9 × 72 = 648

54 × 6 = 324
3 × 86 = 258

Multiplication: Textbook page 21

Assessment

Check-up 10

Name:

876 ÷ 6

```
  6 | 876
     -600    6 × 100
      276
     -240    6 × 40
       36
      -36    6 × 6
        0
```
876 ÷ 6 = 146

825 ÷ 5

```
  5 | 825
     -500    5 × 100
      325
     -300    5 × 60
       25
      -25    5 × 5
        0
```
825 ÷ 5 = 165

619 ÷ 4

```
  4 | 619
     -400    4 × 100
      219
     -200    4 × 50
       19
      -16    4 × 4
        3
```
619 ÷ 4 = 154 r 3

7209 ÷ 9

```
  9 | 7209
     -7200    9 × 800
         9
        -9    9 × 1
         0
```
7209 ÷ 9 = 801

8715 ÷ 7

```
  7 | 8715
     -7000    7 × 1000
      1715
     -1400    7 × 200
       315
      -280    7 × 40
        35
       -35    7 × 5
         0
```
8715 ÷ 7 = 1245

9906 ÷ 8

```
  8 | 9906
     -8000    8 × 1000
      1906
     -1600    8 × 200
       306
      -240    8 × 30
        66
       -64    8 × 8
         2
```
9906 ÷ 8 = 1238 r 2

New Heinemann Maths 6 © DPMC 2002. Copying permitted for purchasing school only. This material is not copyright free.
Division: Textbook page 29

Check-up 9

Name:

56 ÷ 7 = **8**
45 ÷ 5 = **9**
18 ÷ 3 = **6**
72

32 ÷ 8 = **4**
63 ÷ 9 = **7**
1/8 of 64 = **8**
49/7 = 7

42 ÷ 6 = **7**
16 ÷ 2 = **8**
36 ÷ 4 = **9**
40/**5**

28 ÷ 4 = **7**
24 ÷ 3 = **8**
1/6 of 54 = **9**
0 ÷ 10 = 0

37 ÷ 6 = **6 r 1**
83 ÷ 10 = **8 r 3**
29 ÷ 8 = **3 r 5**
1/4 of 35 = **8 r 3**
67 ÷ 7 = **9 r 4**
1/5 of 28 = **5 r 3**

Find one half of each number.

662 → **331**
1780 → **890**
726 → **363**
908 → **454**
1560 → **780**
6200 → **3100**
1940 → **970**
7600 → **3800**

420 ÷ 7 = **60**
270 ÷ **3** = 90
480 ÷ 8 = **60**
350 ÷ **5** = 70
240 ÷ 6 = **40**
210 ÷ **7** = 30

56 ÷ 4 = **14**
108 ÷ 9 = **12**
84 ÷ 7 = **12**

693 ÷ 3 = **231**
824 ÷ 4 = **206**
525 ÷ 5 = **105**
918 ÷ 9 = **102**
606 ÷ 6 = **101**
864 ÷ 8 = **108**

New Heinemann Maths 6 © DPMC 2002. Copying permitted for purchasing school only. This material is not copyright free.
Division: Textbook pages 27–28

Check-up 12

Name:

$\frac{1}{100}$ of 900 = **9**
$\frac{2}{6}$ of 180 = **60**
$\frac{2}{7}$ of 700 = **200**

$\frac{3}{4}$ of 28 = **21**
$\frac{6}{9}$ of 63 = **42**
$\frac{3}{8}$ of 320 = **120**

two fifths of 45 = **18**
five eighths of 56 = **35**
four sixths of 480 = **320**

£$\frac{4}{10}$ → **40 p**
$\frac{1}{3}$ kg → 300 g
$\frac{7}{100}$ km → **70 m**
$\frac{1}{100}$ l → 20 ml
$\frac{8}{100}$ kg → **80 g**
£1 $\frac{37}{100}$ → 137p

$\frac{4}{9}$ of 9 m = **4 m**
$\frac{3}{4}$ of 2 km = **1$\frac{1}{2}$ km**
$\frac{5}{7}$ of 42 kg = **30 kg**
$\frac{3}{5}$ of 4 l = **2$\frac{2}{5}$ l**
$\frac{2}{8}$ of £72 = **£18**
$\frac{7}{10}$ of £5 = **£3·50**

Change to m and mm or km and m or kg and g.

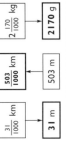
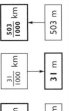
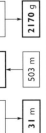

3$\frac{909}{1000}$ km → **3 km 909 m**
1$\frac{429}{1000}$ kg → **1 kg 429 g**
4$\frac{40}{1000}$ m → **4 m 40 mm**
$\frac{6}{1000}$ kg → **0 kg 6 g**

$\frac{264}{1000}$ m → **264 mm**
$\frac{842}{1000}$ m → **842 mm**
$\frac{31}{1000}$ km → **31 m**
$\frac{503}{1000}$ km → **503 m**
2$\frac{170}{1000}$ kg → **2170 g**
5$\frac{8}{1000}$ kg → **5008 g**

New Heinemann Maths 6 © DPMC 2002. Copying permitted for purchasing school only. This material is not copyright free.
Fractions: Textbook pages 44–45

Check-up 11

Name:

Write the equal fractions story for each pair of shapes.

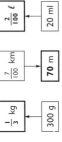

$\frac{18}{30} = \frac{3}{5}$

$\frac{5}{6} = \frac{10}{12}$

$\frac{1}{3} = \frac{4}{12}$
$\frac{3}{4} = \frac{30}{40}$
$\frac{6}{21} = \frac{2}{7}$

$\frac{50}{100} = \frac{1}{2}$
$\frac{16}{40} = \frac{2}{5}$
$\frac{5}{8} = \frac{20}{32}$

$\frac{15}{50} = \frac{3}{10}$
$\frac{5}{7} = \frac{40}{56}$
$\frac{5}{9} = \frac{35}{63}$

$\frac{6}{36} = \frac{1}{6}$
$\frac{1}{9} = \frac{5}{45}$
$\frac{14}{21} = \frac{2}{3}$

Simplify.

$\frac{9}{36}$ → $\frac{3}{12}$
$\frac{49}{70}$ → $\frac{7}{10}$
$\frac{32}{80}$ → $\frac{2}{5}$
$\frac{45}{75}$ → $\frac{9}{15}$

Write True (T) or False (F).

$\frac{1}{7}$ is smaller than $\frac{1}{8}$ — **F**

$\frac{1}{8}$ is twice $\frac{1}{4}$ — **F**

This part of a number line has 30 divisions.
Draw and label arrows to show the positions of $\frac{1}{2}$, $\frac{1}{3}$, $\frac{3}{10}$, $\frac{1}{5}$, $\frac{5}{6}$, $\frac{14}{15}$.

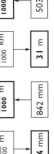
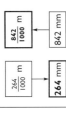

0 — $\frac{1}{10}$ — $\frac{1}{3}$ — $\frac{1}{2}$ — $\frac{3}{5}$ — $\frac{5}{6}$ — $\frac{14}{15}$ — 1

New Heinemann Maths 6 © DPMC 2002. Copying permitted for purchasing school only. This material is not copyright free.
Fractions: Textbook pages 41–43

Check-up 14

Name:

0·24 + 0·53 = **0·77**	0·41 + 0·38 = **0·79**	0·86 + 0·12 = **0·98**
0·5 + 0·67 = **1·17**	0·48 + 0·9 = **1·38**	0·2 + 0·85 = **1·05**

5·14 + 4·25 = **9·39**	1·32 + 4·56 = **5·88**	8·26 + 1·31 = **9·57**
2·3 + 6·47 = **8·77**	4·51 + 4·4 = **8·91**	9·05 + 1·3 = **10·35**

3·82 + **0·18** = 4	**0·55** + 6·45 = 7	8·41 + **0·09** = 8·5
3·57 + **0·03** = 3·6	**4·52** + 1·26 = 5·78	4·17 + **5·51** = 9·68

5·16 + 4·73
```
  5·16
+ 4·73
  9·89
```

7·32 + 4·85
```
  7·32
+ 4·85
 12·17
```

15·51 + 3·24
```
 15·51
+ 3·24
 18·75
```

2·77 + 5·52
```
  2·77
+ 5·52
  8·29
```

13·94 + 8·45
```
 13·94
+ 8·45
 22·39
```

11·34 + 13·63
```
 11·34
+13·63
 24·97
```

Decimals Textbook pages 50–51

Check-up 16

Name:

4·3 × 2 = **8·6**	3 × 2·6 = **7·8**	1·9 × 4 = **7·6**
6 × 2·7 = **16·2**	4·1 × 8 = **32·8**	7 × 4·2 = **29·4**
1·3 × **3** = 3·9	4 × **1·7** = 6·8	6 × **1·3** = 7·8

Double 3·8 **7·6**   Twice 1·9 **3·8**

2·7 × 2 = 5·4   2 × **6·2** = 12·4

Mask, 12·4 kg   Air cylinder, 9·03 kg   Wetsuit, 4·17 kg

Find the total weight of
- 4 masks **49·6 kg**
- 7 air cylinders **63·21 kg**
- 3 wetsuits **12·51 kg**

10 × 2·8 = **28**	6·9 × 10 = **69**	100 × 4·15 = **415**
0·16 × 100 = **16**	100 × 73·8 = **7380**	65·1 × 10 = **651**
10 × 4·9 = 49	0·17 × **100** = 17	100 × **3·09** = 309

6·4 × 30 = **192**	0·56 × 50 = **28**	3·8 × 70 = **266**
60 × 2·9 = **174**	40 × 4·7 = **188**	90 × 0·61 = **54·9**

Decimals Textbook pages 54–56

Check-up 13

Name:

List the first seven numbers in this sequence.

Start at 1.
Count on 0·25 each time.

1 , **1·25** , **1·5** , **1·75** , **2** , **2·25** , **2·5**

Write the 2-place decimal fraction

before 4·35 **4·34**	after 2·99 **3·00**	before 9·71 **9·70**
between 2·88 and 2·90 **2·89**		between 4·11 and 4·09 **4·10**

Which of the numbers has

4·62   0·56   63·2   2·19   9·04   26·4

6 tenths **4·62**	4 hundredths **9·04**	2 units **2·19**
6 tens **63·2**	the smallest tenths digit? **9·04**	

Tick (✓)

- the largest number

1·28   1·82 ✓   1·8   9·65 ✓   9·56   9·6

- the smallest number

5·5   5·74   5·47 ✓   7·62   7·26 ✓   7·6

Round each number

- to the nearest tenth

6·38 ⟶ **6·4**
8·14 ⟶ **8·1**
9·06 ⟶ **9·1**

- to the nearest whole number

14·37 ⟶ **14**
45·81 ⟶ **46**
89·50 ⟶ **90**

Decimals Textbook pages 47–49

Check-up 15

Name:

0·68 − 0·2 = **0·48**	0·93 − 0·7 = **0·23**	0·8 − 0·32 = **0·48**
0·57 − 0·16 = **0·41**	0·75 − 0·34 = **0·41**	0·96 − 0·16 = **0·8**

5·74 − 2·13 = **3·61**	6·85 − 4·11 = **2·74**	7·36 − 2·13 = **5·23**
7·97 − 3·35 = **4·62**	8·58 − 4·26 = **4·32**	9·93 − 7·52 = **2·41**

List pairs of these numbers which have a difference of 0·3.

3·58   0·54   1·14   4·18   0·84   3·88

3·88, 3·58   **0·84, 0·54**   **4·18, 3·88**   **1·14, 0·84**

4·87 − 3·16
```
  4·87
- 3·16
  1·71
```

13·57 − 8·24
```
 13·57
-  8·24
  5·33
```

26·62 − 12·59
```
 26·62
-12·59
 14·03
```

9·47 − 5·81
```
  9·47
- 5·81
  3·66
```

38·25 − 3·69
```
 38·25
-  3·69
 34·56
```

51·06 − 17·38
```
 51·06
-17·38
 33·68
```

Decimals Textbook pages 52–53

Assessment

Check-up 18

Name: _____

Write the decimal fraction shown by each arrow.

2·897 2·920 2·954 3·006

Draw and label arrows to show 2·920 and 3·006.

Write each weight in decimal form.

$\frac{719}{1000}$ kg → 0·719 kg

$7\frac{72}{1000}$ kg → 7·072 kg

9001 thousandths of 1 kg → 9·001 kg

Which of the numbers 5·463 4·536 3·645 6·354 **has**

5 thousandths 3·645
6 hundredths? 5·463
3 tenths 6·354
4 units 4·536

Round to the first decimal place.

0·288 → 0·3   8·049 → 8·0   6·565 → 6·6

Round to the nearest whole number.

27·81 → 27·8   4·477 → 4·5   58·521 → 58·5

1·192 km → 1192 m
0·283 m → 283 mm
9·038 ℓ → 9038 ml
5·130 kg → 5130 g
8·601 km → 8601 m

Decimals: Textbook pages 59–62

Name: _____

Attendances at Superleague grounds

Rocaster Park	41 434	Dunforth Bowl	53 569
Brandvale Field	18 576	Maxford Arena	30 645
Yorkton Stadium	16 052	Northsea Park	7548

1 How many people altogether attended the matches at

- Rocaster and Northsea   49018
- Yorkton and Dunforth   69621
- Brandale and Maxford   49221
- Dunforth and Rocaster?   95003

2 Find the total of each set of numbers.

63 9574 376 → 10013

8 6607 514 72 → 7201

9 684 729 14 507 81 → 16010

Check-up 17

Name: _____

37 ÷ 10 = 3·7   52 ÷ 100 = 0·52   98 ÷ 10 = 9·8

19 ÷ 100 = 0·19   6 ÷ 10 = 0·6   2 ÷ 100 = 0·02

47 ÷ 10 = 4·7   72 ÷ 100 = 0·72   1 ÷ 10 = 0·1

5·6 ÷ 8 = 0·7   4·8 ÷ 6 = 0·8   0·7

3·5 ÷ 7 = 0·5   2·1 ÷ 3 = 0·7   6·3 ÷ 9 = 0·7

4·5 ÷ 5 = 0·9   5

```
13·8 ÷ 6        6 × 1·0
 -6·0           6 × 0·3
  7·8
 -6·0
  1·8
 -1·8
   0
13·8 ÷ 6 = 2·3
```

```
39·6 ÷ 9        9 × 4·0
-36·0           9 × 0·4
  3·6
 -3·6
   0
39·6 ÷ 9 = 4·4
```

```
23 ÷ 5          5 × 4·0
-20·0           5 × 0·6
  3·0
 -3·0
   0
23 ÷ 5 = 4·6
```

```
8·26 ÷ 7        7 × 1·0
 -7·0           7 × 0·1
 1·26           7 × 0·08
 -0·7
 0·56
-0·56
   0
8·26 ÷ 7 = 1·18
```

```
7·48 ÷ 4        4 × 1·0
 -4·0           4 × 0·8
 3·48           4 × 0·07
 -3·2
 0·28
-0·28
   0
7·48 ÷ 4 = 1·87
```

Decimals: Textbook pages 57–58

Name: _____

1 17 + 45 + 23 + 56 = 141   61 + 43 + 66 = 170

Find the total number of beads in the jars.

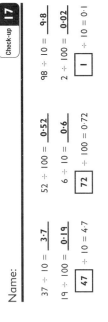

72 71 75 73 78 → 369

2 313 + 229 = 542   676 + 108 = 784
228 + 624 = 852   446 + 435 = 881
519 + 472 = 991   327 + 667 = 994

3 390 + 470 = 860   660 + 150 = 810
686 + 291 = 977   165 + 764 = 929
546 + 382 = 928   394 + 195 = 589

4 Trees in Birdcombe Wood

Birch	Ash	Beech	Elm
730	542	626	815

Find the total number of these trees.

- Elm and Birch 1545
- Beech and Ash 1168
- Ash and Elm 1357
- Birch and Beech 1356

780 + 450 = 1230   736 + 732 = 1468

5 3500 + 4400 = 7900   5600 + 8900 = 14 500
6300 + 3600 = 9900   7800 + 4600 = 12 400

6 3612 + 6177 = 9789   2065 + 4723 = 6788
1058 + 7933 = 8991   5367 + 2414 = 7781

Assessment

Topic Assessment 2b — Subtraction

Name:

1
6900 - 4300 = [2600]   8700 - 4900 = [3800]
940 - [760] = 180   [4500] - 2600 = 1900
14400 - 8800 = [5600]
7800 - 3650 = [4150]   9600 - 2150 = [7450]

2
5467 - 2155 = [3312]   8836 - 4502 = [4334]
3679 - 1456 = [2223]   2598 - 1347 = [1251]
6874 - 359 = [6515]   7982 - 5145 = [2837]

3
4000 - 2597   7000 - 4281   10000 - 3705
= [1403]   = [2719]   = [6295]

4
45397 - 13264   76843 - 32571   58432 - 25706
32133   44272   32726
60000 - 42153   876732 - 56   952381 - 1945
17847   876676   950436

Topic Assessment 2a — Subtraction

Name:

1
520 - 270 = [250]   750 - 470 = [280]   410 - 150 = [260]
940 - [760] = 180   410 - 170 = 140
467 - 190 = [277]   828 - 450 = [378]   235 - 180 = [55]
356 - [270] = 86   634 - 280 = 354
= [310]   = [189]

2
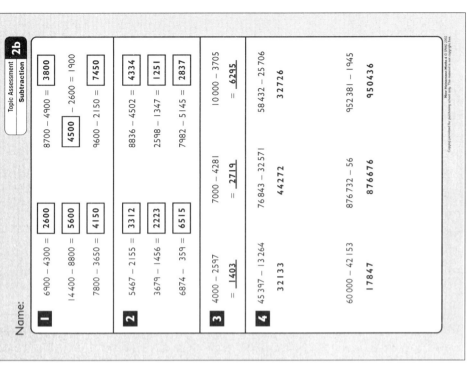
How much heavier is the brick than the tile? [571] g
(tile 389g, brick 960g)

640 - 369 = [271]   750 - 168 = [582]   810 - 626 = [184]
220 - 143 = [77]   460 - 285 = [175]   340 - 151 = [189]

3
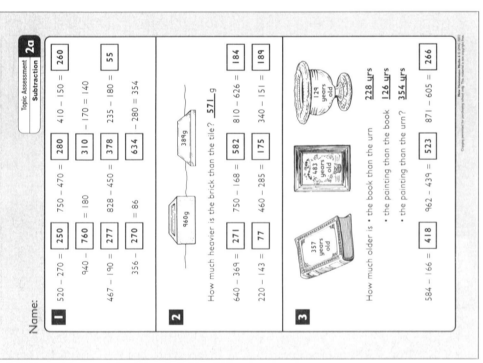
(trophy 129 years old, painting 483 years old, book 357 years old)

How much older is: • the book than the urn?   [228] yrs
                     • the painting than the book   [126] yrs
                     • the painting than the urn?   [354] yrs

584 - 166 = [418]   962 - 439 = [523]   871 - 605 = [266]

Topic Assessment 3b — Multiplication

Name:

1 4 × 1476
4 × 1476 = [5904]

2 5 × 1583
5 × 1583 = [7915]

3 9 × 1047
9 × 1047 = [9423]

4 Find the total cost for 6 adults and 3 children travelling on a *Round the World* cruise.
Round the World CRUISE — Adult £2158, Child £1026
[£16026]

5 26 × 173
26 × 173 = [4498]

6 65 × 149
65 × 149 = [9685]

7 37 × 247
37 × 247 = [9139]

8 Find the cost of 378 concert tickets.
HHO CONCERT 5th May £28
£28 × 378 = [£10584]

Topic Assessment 3a — Multiplication

Name:

1 How much does it cost to buy
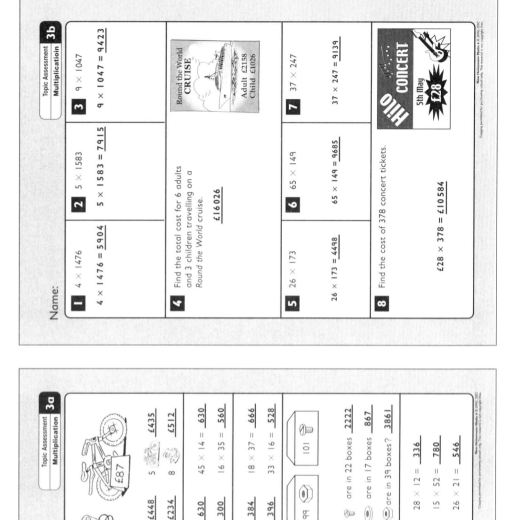
(items: £64, £87, £39)
4   [£156]   5   [£435]
9   [£783]   8   [£512]
7   [£448]   6   [£234]

2
15 × 14 = [210]   35 × 18 = [630]   45 × 14 = [630]
74 × 5 = [370]   12 × 25 = [300]   16 × 35 = [560]

3
12 × 23 = [276]   16 × 24 = [384]   18 × 37 = [666]
41 × 14 = [574]   22 × 18 = [396]   33 × 16 = [528]

4 How many
(screw 51, bolt 49, nut 99, bolt 101)
are in 18 boxes [1782]   are in 22 boxes [2222]
are in 23 boxes [1127]   are in 17 boxes [867]
are in 36 boxes [1836]   are in 39 boxes? [3861]

5
14 × 37 = [518]   28 × 12 = [336]
101 × 19 = [1919]   15 × 52 = [780]
24 × 13 = [312]   26 × 21 = [546]

Assessment

Name: _____

1

$912 \div 8$

```
   8 ) 9 1 2
     - 8 0 0      8 × 100
       1 1 2
     -   8 0      8 × 10
         3 2
     -   3 2      8 × 4
           0
```

$912 \div 8 = 114$

$646 \div 6$

```
   6 ) 6 4 6
     - 6 0 0      6 × 100
         4 6
     -   4 2      6 × 7
           4
```

$646 \div 6 = 107 \text{ r } 4$

$793 \div 9$

```
   9 ) 7 9 3
     - 7 2 0      9 × 80
         7 3
     -   7 2      9 × 8
           1
```

$793 \div 9 = 88 \text{ r } 1$

2 Share equally

• 1595 desks among 5 offices

```
   5 ) 1 5 9 5
     - 1 5 0 0    5 × 300
           9 5
     -     5 0    5 × 10
           4 5
     -     4 5    5 × 9
             0
```

$1595 \div 5 = 319$

• 3732 chairs among 6 schools.

```
   6 ) 3 7 3 2
     - 3 6 0 0    6 × 600
         1 3 2
     -     7 2    6 × 12
           6 0
     -     6 0    6 × 10
             0
```

$3732 \div 6 = 622$

New Heinemann Maths 6 © DPHC 2002
Copying permitted for purchasing school only. This material is not copyright free.

Name: _____

1 Write this number sequence.

Start at 48. Count back in 7s to 6.

48, **41**, **34**, **27**, **20**, **13**, **6**

2 Tick (✓) the rule for this number sequence → 1, 2, 4, 8, 16, 32, 64

Add 2 each time

Multiply by 2 each time ✓

Divide by 2 each time

3 Complete each number sequence and write its rule.

44, **63**, **82**, **101**, **120**, **139** → **Add 19**

87, **72**, **57**, **42**, **27**, **12** → **Subtract 15**

4 Which of the numbers on the cards are

49   18   36   10   30   25   21

square numbers **49, 36, 25**

triangular numbers **36, 10, 21**

even and **not** triangular? **18, 30**

square **and** odd **49, 25**

5 What type of number is the product of two odd numbers?

An odd number

6 What is the new temperature after a temperature of -9°C

rises by 5 degrees **-4°C**

falls by 5 degrees **-14°C**

rises by 9 degrees **9°C**

falls by 9 degrees **-18°C**

7 Find the difference, in degrees, between temperatures of

3°C and 12°C **9°**

-3°C and -12°C **9°**

3°C and -12°C **15°**

New Heinemann Maths 6 © DPHC 2002
Copying permitted for purchasing school only. This material is not copyright free.

Name: _____

1

$48 \div 8 = \boxed{6}$ $\frac{1}{6}$ of $60 = \boxed{10}$

$36 \div 9 = \boxed{4}$ $\frac{1}{4}$ of $0 = \boxed{0}$

$54 \div 9 = \boxed{6}$ $28 \div 7 = \boxed{4}$

$100 \div 10 = \boxed{10}$ $48 \div 6 = \boxed{8}$

$18 \div 2 = 9$ $90 \div \boxed{10} = 9$

$\boxed{45} \div 9 = 5$ $16 \div 8 = 2$

2

$48 \div 9 = \mathbf{5 \text{ r } 3}$ $42 \div 10 = \mathbf{4 \text{ r } 2}$

$33 \div 7 = \mathbf{4 \text{ r } 5}$ $27 \div 6 = \mathbf{4 \text{ r } 3}$

$38 \div 5 = \mathbf{7 \text{ r } 3}$

$76 \div 8 = \mathbf{9 \text{ r } 4}$

3 Find one half of each number.

494 → **247**   706 → **353**

386 → **193**   4760 → **2380**

1140 → **570**

5200 → **2600**

4

$450 \div 9 = \boxed{50}$ $320 \div 4 = \boxed{80}$

$540 \div 6 = \boxed{90}$ $480 \div \boxed{8} = 60$

$630 \div \boxed{9} = 70$

$250 \div 5 = \boxed{50}$

5

$52 \div 4 = \boxed{13}$

$72 \div 6 = \boxed{12}$

$98 \div 7 = \boxed{14}$

6

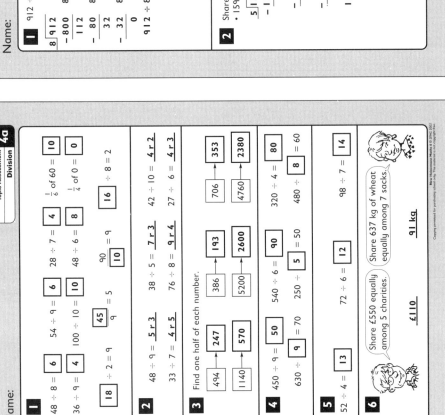
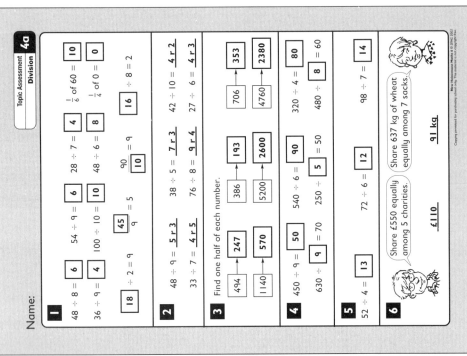

Share £550 equally among 5 charities. **£110**

Share 637 kg of wheat equally among 7 sacks. **91 kg**

New Heinemann Maths 6 © DPHC 2002
Copying permitted for purchasing school only. This material is not copyright free.

Name: _____

1

```
  33 ) 3 6 3
     - 3 3 0      33 × 10
         3 3
     -   3 3      33 × 1
           0
```

$363 \div 33 = 11$

2

```
  26 ) 5 9 8
     - 5 2 0      26 × 20
         7 8
     -   7 8      26 × 3
           0
```

$598 \div 26 = 23$

3

```
  21 ) 6 7 2
     - 6 3 0      21 × 30
         4 2
     -   4 2      21 × 2
           0
```

$672 \div 21 = 32$

4

```
  22 ) 6 3 8
     - 4 4 0      22 × 20
         1 9 8
     - 1 1 0      22 × 5
           8 8
     -     8 8    22 × 4
             0
```

$638 \div 22 = 29$

5

```
  16 ) 5 1 5
     - 4 8 0      16 × 30
         3 5
     -   3 2      16 × 2
           3
```

$515 \div 16 = 32 \text{ r } 3$

6

```
  17 ) 7 5 1
     - 5 1 0      17 × 30
         2 4 1
     - 1 7 0      17 × 10
           7 1
     -     6 8    17 × 4
             3
```

$751 \div 17 = 44 \text{ r } 3$

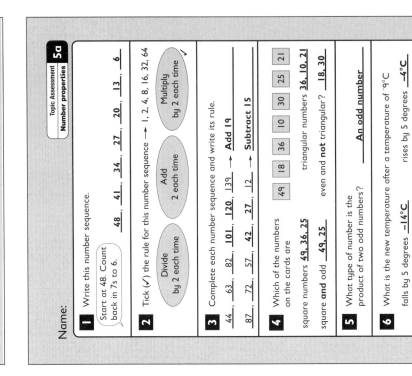

Assessment

Topic Assessment 6a — Fractions

Name:

1 Change to improper fractions or mixed numbers.

$2\frac{7}{9}$ $\frac{25}{9}$ $\frac{38}{5}$ $7\frac{3}{5}$ $5\frac{3}{8}$ $\frac{43}{8}$ $\frac{62}{15}$ $4\frac{2}{15}$

2 Shade the second shape in each pair to match the first one.
Complete each equal fractions story.

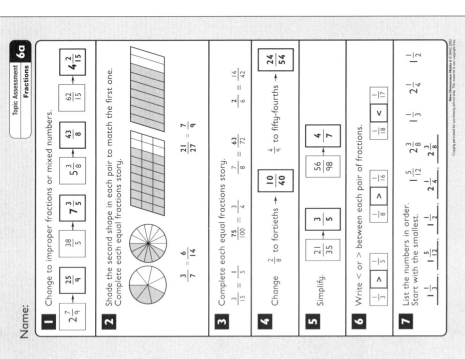

$\frac{3}{7} = \frac{6}{14}$ $\frac{21}{27} = \frac{7}{9}$

3 Complete each equal fractions story.

$\frac{3}{15} = \frac{1}{5}$ $\frac{75}{100} = \frac{3}{4}$ $\frac{63}{72} = \frac{7}{8}$

4 Change $\frac{2}{8}$ to fortieths $\frac{10}{40}$ $\frac{2}{6} = \frac{14}{42}$ $\frac{4}{9}$ to fifty-fourths $\frac{24}{54}$

5 Simplify. $\frac{21}{35}$ $\frac{3}{5}$ $\frac{56}{98}$ $\frac{4}{7}$

6 Write < or > between each pair of fractions.

$\frac{1}{8}$ > $\frac{1}{5}$ $\frac{1}{18}$ < $\frac{1}{17}$

7 List the numbers in order. Start with the smallest.

$1\frac{5}{12}$ $2\frac{3}{8}$ $1\frac{1}{3}$ $2\frac{1}{4}$

$1\frac{1}{3}$, $1\frac{5}{12}$, $2\frac{1}{4}$, $2\frac{3}{8}$

Topic Assessment 7a — Decimals

Name:

1 Complete each sequence.

3·94, 3·96, 3·98, __4__ , __4·02__ , __4·04__ , 4·06

1·05, 1·03, 1·01, __0·99__ , __0·97__ , __0·95__ , 0·93

2 Write each decimal as a mixed number.

7·93 ⟶ $7\frac{93}{100}$ 32·7 ⟶ $32\frac{7}{10}$

13·5 ⟶ $13\frac{5}{10}$ 1·09 ⟶ $1\frac{9}{100}$

3 Write the value of each circled digit.

9·8̲3 __three hundredths__   2̲0·8 __zero units__

5̲5·2 __five tens__   0·3̲7 __three tenths__

4 Write the numbers in order. Start with the smallest.

7·62 6·7 7·26 7·0 6·27

__6·27__ , __6·7__ , __6·72__ , __7·0__ , __7·26__ , __7·62__

5 0·88 + 0·6 = __1·48__   1·25 + 3·71 = __4·96__   4·3 + 1·68 = __5·98__

6
```
  2·42 + 4·83
    4·83
  + 2·42
    7·25
```

7
```
  9·58 + 7·76
    9·58
  + 7·76
   17·34
```

8 0·68 − 0·4 = __0·28__   8·78 − 3·46 = __5·32__   5·8 − 1·62 = __4·18__

9
```
  58·62 − 23·14
   5 8·⁵6̸²
  − 2 3·1 4
   3 5·4 8
```

10
```
  64·43 − 22·76
   6⁶⁴·¹³4̸³
  − 2 2·7 6
   4 1·6 7
```

Topic Assessment 5b — Number properties

Name:

1 How can you tell, **without dividing by 8**, that 376 is exactly divisible by 8? __Add the digits of the number together. If that number is divisible by 8 so is the original number.__

2 How can you tell, **without dividing by 6**, that 138 is exactly divisible by 6? __Add the digits of the number together. If that number is divisible by 6 so is the original number.__

3 List the numbers between 60 and 80 which are multiples of

8 __64, 72, 80__   9 __63, 72__   8 and 9 __72__

4 What is the smallest number that is a common multiple of

3 and 5 __15__   10 and 6 __30__   7 and 28? __28__

5 For the number **30**, list all the

factor pairs   __(1, 30) (2, 15) (3, 10) (5, 6)__

prime factors. __2, 3, 5__

6 Write a multiplication with a product of 20. Use only prime factors of 20.   __2 × 2 × 5__

7 Write True (T) or False (F) for each statement about 36.

36 is an odd number. __F__   36 is a multiple of 12. __T__

36 is a common multiple of both 9 and 4. __F__   36 is a triangular number. __T__

8 is a factor of 36. __F__   36 is **not** a square number. __F__

Topic Assessment 6b — Fractions

Name:

1 $\frac{1}{10}$ of 80 = __8__ $\frac{2}{3}$ of 27 = __18__   six sevenths of 35 = __30__

$\frac{2}{5}$ of 500 = __200__ $\frac{3}{6}$ of 540 = __270__   eight ninths of 270 = __240__

2 $\frac{6}{10}$ kg __600 g__ $\frac{43}{100}$ £ __43 p__ $\frac{8}{100}$ ℓ __80 ml__

3 $\frac{3}{4}$ of £24 = __£18__ $\frac{8}{10}$ of 3 km = $2\frac{2}{5}$ km $\frac{3}{8}$ of 6 m = $2\frac{1}{4}$ m

4 $\frac{2·280}{1·000}$ km __2280 m__ $\frac{616}{1000}$ kg __616 g__ $3\frac{52}{1000}$ ℓ __3052 ml__

5 Change $7\frac{831}{1000}$ m to m and mm. __7 m 831 mm__

Change $5\frac{4}{1000}$ kg to kg and g. __5 kg 4 g__

6 Write True (T) or False (F) for each statement about the design.

• There is 1 shaded square to every 5 unshaded squares. __T__

• 5 in every 6 squares are unshaded. __T__

• The proportion of shaded squares is $\frac{1}{5}$ of all the squares. __F__

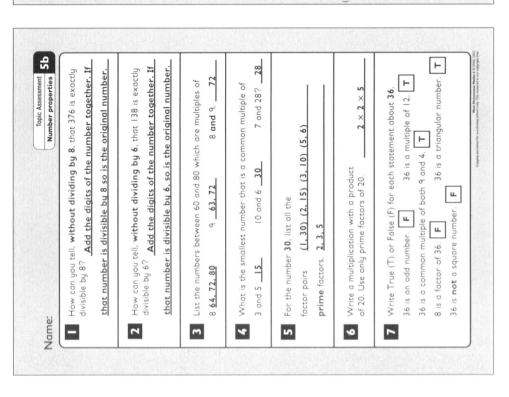

7 In the design,

• how many **shaded to unshaded** squares are there   __1 to 3__

• what fraction is the number of shaded squares of the number of unshaded squares   __$\frac{1}{4}$__

• what proportion of **all** the squares is unshaded?   __$\frac{3}{4}$__

Assessment

Topic Assessment 7c — Decimals

Name:

1 Write as a decimal. $\frac{931}{1000} \rightarrow$ **0·931** $6\frac{4}{1000} \rightarrow$ **6·004**

Write as a fraction. 3·749 $\rightarrow$ **$3\frac{749}{1000}$**   8·060 $\rightarrow$ **$8\frac{6}{100}$**

2 Complete each sequence.
1·092, 1·094, 1·096, **1·098**, **1·100**, **1·102**, 1·104
4·618, 4·612, 4·606, **4·600**, **4·594**, **4·588**, 4·582

3 Write each number as a 3-place decimal fraction.
two tenths, zero hundredths and six thousandths **0·206**
nine and twenty-seven thousandths **9·0027**

4 What is the value of the digit in each circle?
7·③19   **3 tenths**   8·91⑦   **7 thousandths**

5 Tick (✓) the larger number.
5·421✓ or 5·412
9·009 or 9·090✓
$6\frac{80}{1000}$ or 6·8✓

6 Write each set of numbers in order, starting with the smallest.
3·354, 4·553, 4·543, 3·353, 3·345   **3·345, 3·353, 3·354, 4·543, 4·553**
7·867, 8·678, 8·67, 7·8, 8·6   **7·8, 7·867, 8·6, 8·67, 8·678**

7 Round to the first decimal place.
2·286 $\rightarrow$ **2·3**   0·705 $\rightarrow$ **0·7**   1·051 $\rightarrow$ **1·1**

8 Find the **approximate** total of 36·726 and 45·432 after rounding each to the nearest whole number. **82**

9 A drink is made by mixing 375 ml of orange juice and 1·5 ℓ of water. Find, in decimal form, the total volume of the mixture. **1·875 ℓ**

Topic Assessment 7b — Decimals

Name:

1 3 × 3·4 = **10·2**   8 × 5·6 = **44·8**   7 × 6·3 = **44·1**
5 × [**1·7**] = 8·5   4·7 × [**4**] = 18·8

2 Find the total length of
• 6 ropes.   [16·9m long]
• 8 poles.   [2·37m long]
6 × 16·9 m = **101·4 m**   8 × 2·37 m = **18·96 m**

3 40 × 1·6 = **64**   60 × 3·7 = **222**   2·4 × 80 = **192**
0·89 × 20 = **17·8**   50 × 0·07 = **3·5**   90 × 5·4 = **486**

4 46 ÷ 10 = **4·6**   38 ÷ 100 = **0·38**   3 ÷ 10 = **0·3**
[**61**] ÷ 100 = 0·61   [**5·4**] ÷ 10 = 0·54

5 ½ of 0·36 = **0·18**   half of 1·6 = [**1·6**] = 0·8

6 2·4 ÷ 3 = **0·8**   7·2 ÷ 8 = **0·9**   1·2 ÷ 6 = **0·2**
[**3·2**] ÷ 4 = 0·8   6·3 ÷ 9 = 0·7

7
```
   25·8 ÷ 6
 6 )25·8      6 × 4·0
   -24        6 × 0·2
    1·8       6 × 0·3
   -1·8
     0
 25·8 ÷ 6 = 4·3
```

8
```
   9·84 ÷ 8
 8 )9·84      8 × 1·0
   -8         8 × 0·2
    1·84      8 × 0·03
   -1·6
    0·24
   -0·24
     0
 9·84 ÷ 8 = 1·23
```

Topic Assessment 8b — Percentages

Name:

1 Write as a fraction **and** as a percentage.
53 out of 100   **$\frac{53}{100}$**   **53%**
45 out of 90   **$\frac{1}{2}$**   **50%**
200 out of 800   **$\frac{1}{4}$**   **25%**
100 out of 1000   **$\frac{1}{10}$**   **10%**

2 Find
40% of 50   **20**
80% of 60   **48**
100% of 300.   **300**

3 Write as a percentage.
6 out of 60   **10%**
18 out of 30   **60%**
28 out of 56.   **50%**

4 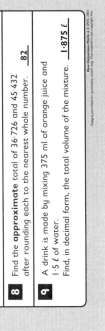 60% of this 2 metre anchor chain is rusted.
What length of Alf's chain is rusted?   **1·2 m**

5 Find • 15% of 80   **12**   • 12½% of 1200.   **150**

6 Write in two other ways.
79%   **0·79**   **$\frac{79}{100}$**
0·18   **18%**   **$\frac{18}{100}$**
3%   **0·03**   **$\frac{3}{100}$**
$\frac{3}{20}$ **15%**   **0·15**

7 Write in order, starting with the **smallest** number.
7   0·7   77% $\frac{7}{100}$ 77
$\frac{7}{100}$, 0·7, 77%, 7, 77

Topic Assessment 8a — Percentages

Name:

1 Colour.
57%   25%   60%   75%   90%

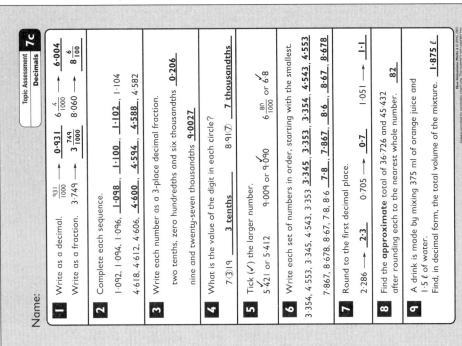

2 For each shape, write the amount shaded as a fraction **and** as a percentage.
$\frac{6}{20}$ **30%**
$\frac{5}{25}$ **20%**
$\frac{9}{10}$ **90%**
$\frac{15}{30}$ **50%**

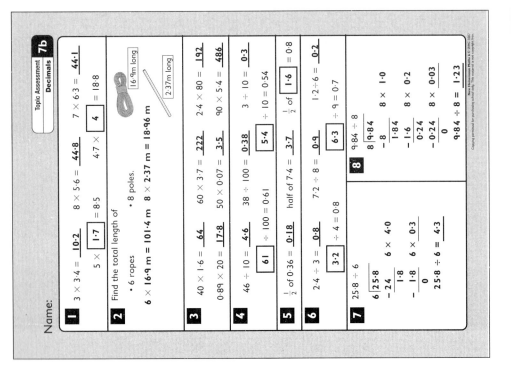
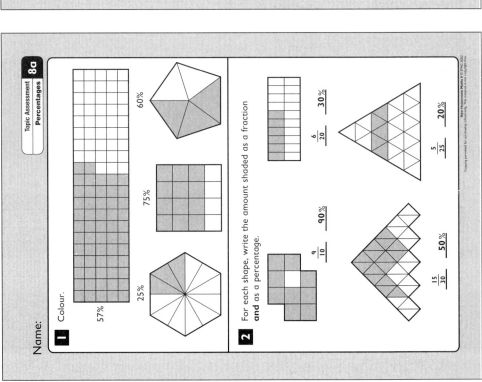

Assessment

Topic Assessment 10 — Length

Name:

1 Each model plane is drawn to a different scale. Measure each plane, then calculate its true length.

Scale: 1 cm to 4 cm   Scale: 1 cm to 20 cm

True length = **16 cm**   True length = **120 cm**

2 This plan of a car park is drawn to a scale of 1 cm to 5 m. Find the true length and breadth of the car park.

True length = **35 cm**
True breadth = **25 cm**

3 Complete.

cm	3.7 cm	6.4 cm	2.9 cm	5.3 cm	7.2 cm
mm	37 mm	64 mm	29 mm	53 mm	72 mm
cm and mm	3 cm 7 mm	6 cm 4 mm	2 cm 9 mm	5 cm 3 mm	7 cm 2 mm

4 Find the distance, in metres, between

Lugton $1\frac{3}{10}$ km Brem 2 km 450 m Tipvale $2\frac{1}{2}$ km Gyle

Lugton and Tipvale **3750 m**
Brem and Gyle **5200 m**

5 Calculate the perimeter of each shape.

5 m 7 m 4 m 1 m   Perimeter = **32 m**

24 m 8 m 1 m   Perimeter = **64 m**

Topic Assessment 9 — Weight

Name:

1 Write each weight, in grams, to the nearest 100 g.

2100 g   **4800 g**

2 Write each weight, in kilograms, to the nearest $\frac{1}{10}$ kg.

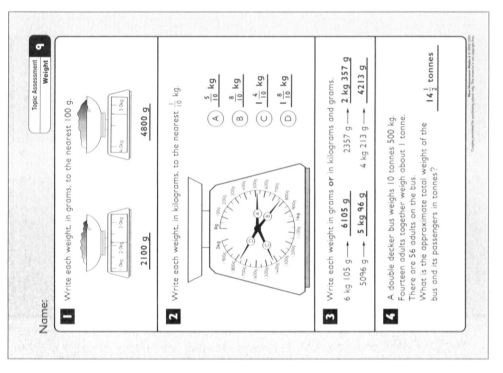

A) $\frac{5}{10}$ kg
B) $\frac{8}{10}$ kg
C) $1\frac{4}{10}$ kg
D) $1\frac{8}{10}$ kg

3 Write each weight in grams **or** in kilograms and grams.

6 kg 105 g → **6105 g**   2357 g → **2 kg 357 g**
5096 g → **5 kg 96 g**   4 kg 213 g → **4213 g**

4 A double decker bus weighs 10 tonnes 500 kg. Fourteen adults together weigh about 1 tonne. There are 56 adults on the bus. What is the approximate total weight of the bus and its passengers in tonnes?

$14\frac{1}{2}$ tonnes

Topic Assessment 11b — Time

Name:

1 Write the time.

• 25 min after 07.50 **08.15**
• 1 h and 35 min after 23.10 **00.45**
• 55 min before 10.15 **09.20**
• 1 h and 45 min earlier than 01.25 **23.40**

2 How many minutes are there between each **start** and **finish** time?

start 07:20 finish 08:10 **50 min**
start 10:25 finish 11:40 **75 min**
start 15:45 finish 17:35 **110 min**
start 22:55 finish 00:50 **115 min**

3 **Cinema-scope**

Marooned (lasts 115 min)
Start times: 1100, 1315, 1530, 1805, 2035

Rescued (lasts 105 min)
Start times: 1000, 1155, 1305, 1515, 1750, 2015

Two friends arrive at **Cinema-scope** at 1030. They watch **both** films.

What is the **earliest** time they could finish watching both films? **1450**

Explain. **They go to the 1100 showing of Marooned which finishes at 1255, then the 1305 showing of Rescued, finishing at 1450.**

Topic Assessment 11a — Time

Name:

1 Write each time as a 24-hour time.

3:00 AM → **03:00**
1:27 PM → **13:27**
10:06 PM → **22:06**

2 Write each time as a 12-hour time. Use am or pm.

06:30 → **6.30 am**
15:33 → **3.33 pm**
18:09 → **6.09 pm**

3 Write the times in order. Start with the earliest.

9.12 pm   19:24   noon   12:03   21:21   7.42 pm

noon   12:03   19:24   7.42 pm   9.12 pm   21:21

4 Match the times.

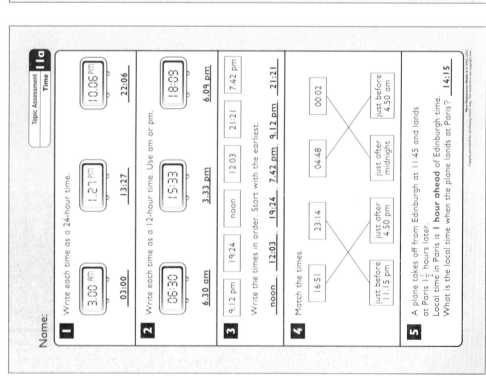

16:51   23:14   04:48   00:02

just before 11.15 pm   just after 4.50 pm   just after midnight   just before 4.50 am

5 A plane takes off from Edinburgh at 11.45 and lands at Paris at 11.45 and lands at Paris $1\frac{1}{2}$ hours later. Local time in Paris is **1 hour ahead** of Edinburgh time. What is the local time when the plane lands at Paris? **14:15**

Assessment

58

Topic Assessment 13 — Volume

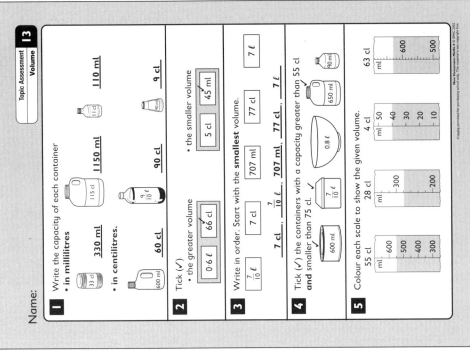

Name:

1 Write the capacity of each container
- **in millilitres**
- **in centilitres.**

330 ml   1150 ml   110 ml

60 cl   90 cl   9 cl

2 Tick (✓)
- the greater volume
- the smaller volume

0·6 ℓ   66 cl

5 cl   45 ml

3 Write in order. Start with the **smallest** volume.

7 cl   707 ml   77 cl   7 ℓ

7 cl , $\frac{7}{10}$ ℓ , 707 ml , 77 cl , 7 ℓ

4 Tick (✓) the containers with a capacity greater than 55 cl **and** smaller than 75 cl.

$\frac{7}{10}$ ℓ   0·8 ℓ $\frac{7}{10}$ ℓ ✓   650 ml

5 Colour each scale to show the given volume.

55 cl   28 cl   4 cl   63 cl

New Heinemann Maths 4 © SPMG 2002
Copying permitted for purchasing school only. This material is not copyright free.

Topic Assessment 14b — 2D shape

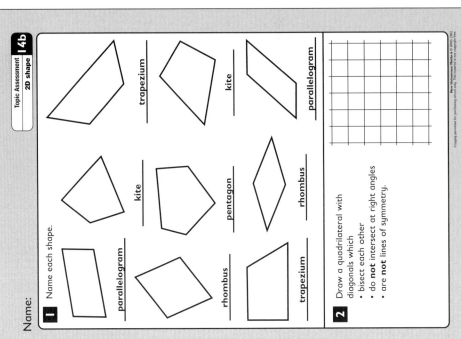

Name:

1 Name each shape.

parallelogram   kite   trapezium

rhombus   pentagon   kite

trapezium   rhombus   parallelogram

2 Draw a quadrilateral with diagonals which
- bisect each other
- do **not** intersect at right angles
- are **not** lines of symmetry.

New Heinemann Maths 4 © SPMG 2002
Copying permitted for purchasing school only. This material is not copyright free.

Topic Assessment 12 — Area

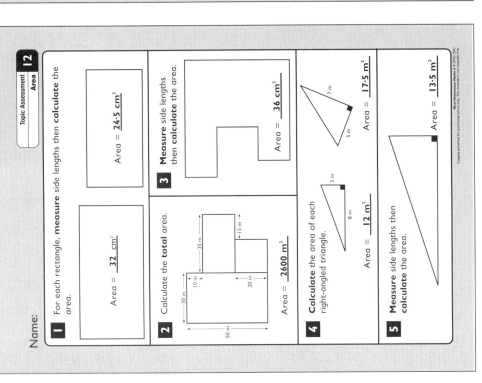

Name:

1 For each rectangle, **measure** side lengths then **calculate** the area.

Area = 32 cm²   Area = 24·5 cm²

2 Calculate the **total** area.

30 m   35 m   15 m
50 m   10 m   20 m

Area = 2600 m²

3 **Measure** side lengths then **calculate** the area.

Area = 36 cm²

4 **Calculate** the area of each right-angled triangle.

3 m   8 m   Area = 12 m²

7 m   5 m   Area = 17·5 m²

5 **Measure** side lengths then **calculate** the area.

Area = 13·5 m²

New Heinemann Maths 4 © SPMG 2002
Copying permitted for purchasing school only. This material is not copyright free.

Topic Assessment 14a — 2D shape

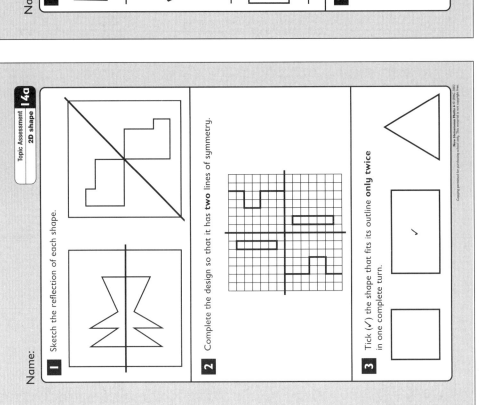

Name:

1 Sketch the reflection of each shape.

2 Complete the design so that it has **two** lines of symmetry.

3 Tick (✓) the shape that fits its outline **only twice** in one complete turn.

✓

New Heinemann Maths 4 © SPMG 2002
Copying permitted for purchasing school only. This material is not copyright free.

Topic Assessment 15b — Position, movement, angle

Name:

1 For each angle, write acute or obtuse or reflex.

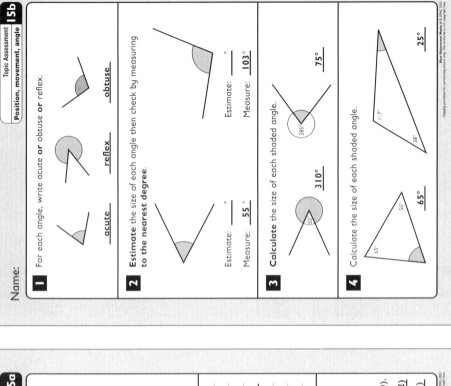

acute   reflex   obtuse

2 Estimate the size of each angle then check by measuring to the nearest degree.

Estimate: ° Measure: 55°

Estimate: ° Measure: 103°

3 Calculate the size of each shaded angle.

310°   75°

4 Calculate the size of each shaded angle.

65°   25°

Round-up

Name:

1
$63 \times 100 = \underline{6300}$ $142 \times 1000 = \underline{142\,000}$

$81\,000 \div 1000 = \underline{81}$ $4500 \div 100 = \underline{45}$

2
$486 + 270 = \underline{756}$ $637 + 156 = \underline{793}$

$4635 + 2104 = \underline{6739}$ $6014 + 2867 = \underline{8881}$

3

Car	Price
Ford	£7896
Renault	£5274
Rover	£6493
BMW	£8000

Find the difference between the price of the
- Ford and the Renault   £2622
- Rover and the Renault   £1219
- Rover and the BMW.   £1507

4
$46\,251 - 12\,517$
$= 33\,734$

$783\,265 - 6927$
$= 776\,338$

5
$45 \times 14 = \underline{630}$ $18 \times 26 = \underline{468}$ $50 \times 48 = \underline{2400}$

6
34×186
$= 6324$

27×235
$= 6345$

Topic Assessment 15a — Position, movement, angle

Name:

1 What are the co-ordinates of the

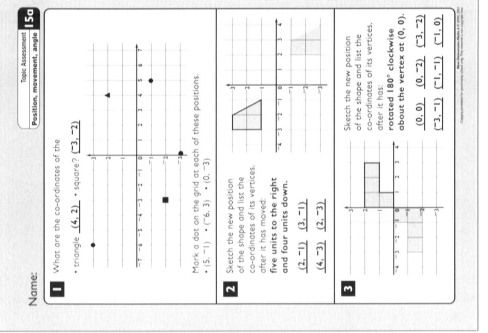

- triangle (4, 2) • square? (⁻3, ⁻2)
- (5, ⁻1) • (6, 3) • (0, ⁻3)

2 Mark a dot on the grid at each of these positions.

Sketch the new position of the shape and list the co-ordinates of its vertices, after it has moved:
five units to the right and four units down.

(2, ⁻1) (3, ⁻1)
(4, ⁻3) (2, ⁻3)

3 Sketch the new position of the shape and list the co-ordinates of its vertices, after it has:
rotated 180° clockwise about the vertex at (0, 0).

(0, 0) (0, ⁻2) (⁻3, ⁻2)
(⁻3, ⁻1) (⁻1, ⁻1) (⁻1, 0)

Topic Assessment 16 — Data handling

Name:

1 These are the points scored by 9 children in a quiz.

46 48 44 46 45 43 46 44 43

Write the points scored in order.

43, 43, 44, 44, 45, 46, 46, 46, 48

Find • the range 5 • the mode 46 • the median. 45

2 Some Year 6 children found out how long they took to walk to school. Their times are shown in the table.

Year 6 Walking Times (Minutes)			
12	11	9	7
10	11	9	12

Find the **total** walking time of all the children.   70 min

What was the mean walking time?   10 min

3 The graph shows the number of lengths swum by **60** children in their sponsored swim.

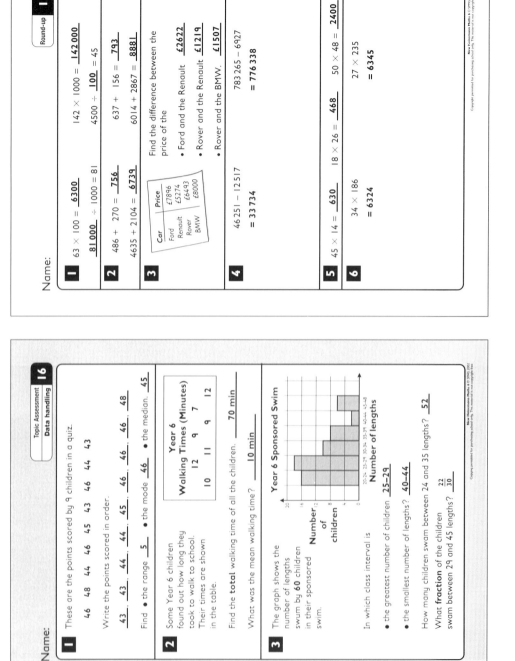

Year 6 Sponsored Swim

Number of children / Number of lengths

20-24 25-29 30-34 35-39 40-44 45-48

In which class interval is
- the greatest number of children 25-29
- the smallest number of lengths? 40-44

How many children swam between 24 and 35 lengths? 52

What **fraction** of the children swam between 29 and 45 lengths? $\frac{22}{30}$

Top-left panel

Name: _____   **Round-up** 1

13 Write 893 g in thousandths of 1 kg → $\frac{893}{1000}$ **kg**

Change $3\frac{145}{1000}$ km to km and m → **3 km 145 m**

14 0·43 + 0·36 = **0·79**   0·7 + 0·91 = **1·61**   3·5 + 6·42 = **9·92**

15 3·68 + 5·73   15·76 + 8·39
= **9·41**   = **24·15**

16 90 × 1·3 = **117**   0·87 × 30 = **26·1**   60 × 0·24 = **14·4**

17 43·2 ÷ 9   6·44 ÷ 7
= **4·8**   = **0·92**

18 Round to the
• first decimal place   3·739 → **3·7**
• nearest whole number. 7·499 → **7**

19 Write as a fraction **and** as a percentage.
17 out of 100 $\frac{17}{100}$ **17%**   300 out of 400 $\frac{3}{4}$ **75%**

20 Write 73% as a fraction **and** as a decimal. $\frac{73}{100}$ **0·73**
Write 0·06 as a fraction **and** as a percentage. $\frac{6}{100}$ **6%**

Top-right panel

Name: _____   **Round-up** 2

1 Increase 4 217 615 by 10 000.   **4 227 615**
Decrease 6 975 201 by 100 000.   **6 875 201**

2 Round to the nearest
• hundred 861   **900**   • hundred 5249   **5200**
• thousand 172 136   **172 000**   • million 18 500 007   **19 000 000**

3 560 + 820 = **1380**   630 + 725 = **1355**   421 + 958 = **1379**
660 + **930** = 1590   **837** + 550 = 1387

4 53 281 + 16 565   33 604 + 5876
= **69 846**   = **39 480**

5 Find the difference between
920 and 463   **457**   742 and 327   **415**

6 Find the cost of
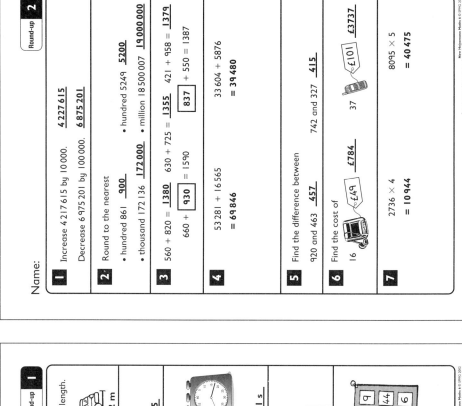
16 [£49]   **£784**   37 [£101]   **£3737**

7 2736 × 4   8095 × 5
= **10 944**   = **40 475**

Bottom-left panel

Name: _____   **Round-up** 1

7 360 ÷ 4 = **90**   630 ÷ **9** = 70   98 ÷ 7 = **14**
51 ÷ 3 = **17**   618 ÷ 6 = **103**   427 ÷ 7 = **61**

8 23 | 552   18 | 653
552 ÷ 23 = **24**   653 ÷ 18 = **36 r 5**

9 Complete each number sequence and write the rule.
17, 38, 59, **80**, **101**, **122** →   **Add 21**
97, **84**, 71, 58, **45**, 32 →   **Subtract 13**

10 Find, in degrees, the difference between temperatures of
5°C and ⁻8°C   **13°**   ⁻4°C and ⁻12°C.   **8°**

11 For the number 24, list all the
• factor pairs   **1, 24; 2, 12; 3, 8; 4, 6**
• **prime** factors.   **2, 3**

12 List the numbers in order. $1\frac{4}{20}$, $1\frac{3}{5}$, $2\frac{3}{10}$, $2\frac{1}{4}$, $1\frac{1}{2}$
Start with the largest.
$2\frac{3}{10}$ $2\frac{1}{4}$ $1\frac{3}{5}$ $1\frac{1}{2}$ $1\frac{4}{20}$

Bottom-right panel

Name: _____   **Round-up** 1

21 Measure the truck in centimetres then calculate its true length.
Scale: 1 cm to 2 m
True length = **10 m**

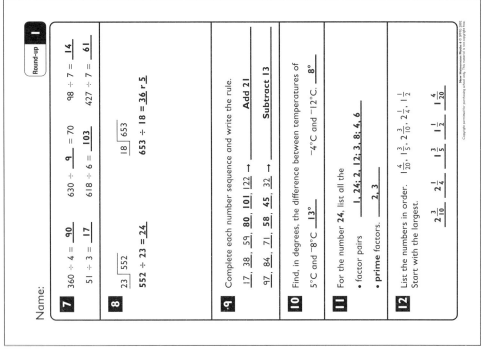

22 Write the time.
• 1 hr and 45 min after 09.40   **11.25**
• 80 minutes before 13.05.   **11.45**

23 How long was spent on each activity?
10 min 25 s   **17 min 41 s**

24 Tick (✓)
• the greater volume   • the smaller volume
$\frac{1}{2}\ell$ | 7 cl   4 cl | 35 ml

25 List **all** the possible outcomes if the dart hits a card with

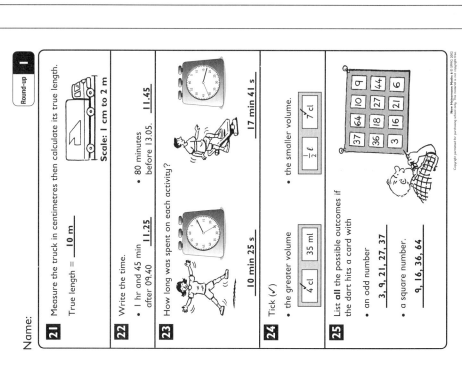

• an odd number   **3, 9, 21, 27, 37**
• a square number.   **9, 16, 36, 64**

Assessment

Round-up 2

Name:

8 Find an approximate answer for each division.
- $794 \div 18 \rightarrow$ **800** $\div$ **20** $=$ **40**
- $616 \div 32 \rightarrow$ **600** $\div$ **30** $=$ **20**

9

| 5 | 81 | 9 | 100 | 21 | 30 | 6 |

Which of the numbers are
- square numbers **9, 81, 100**
- square and odd **9, 81**
- multiples of 5 **5, 30, 100**
- triangular numbers? **6, 21**

10 List the first six multiples of 8. **8, 16, 24, 32, 40, 48**
Write the smallest number that is a **common** multiple of 5 and 9. **45**

11 Simplify.
$\dfrac{18}{24} = \dfrac{3}{4}$ $\dfrac{56}{80} = \dfrac{7}{10}$ $\dfrac{60}{75} = \dfrac{4}{5}$

12 $\frac{1}{6}$ of $120 =$ **20** $\frac{4}{9}$ of $81 =$ **36** $\frac{7}{10}$ of 2 km $=$ **1·4 km**

13 Round to the nearest tenth.
$0·63 \rightarrow$ **0·6** $9·09 \rightarrow$ **9·1**
Round each length to the nearest whole metre.
$28·67$ m $\rightarrow$ **29 m** $99·08$ m $\rightarrow$ **99 m**

14 $6·85 - 4·23 =$ **2·62** $9·57 - 5·4 =$ **4·17** $7·98 - 2·67 =$ **5·31**

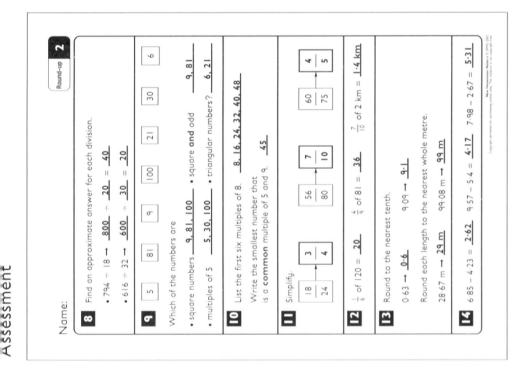

Round-up 2

Name:

15 $7·68 - 2·42$
$= 5·26$

$93·42 - 47·84$
$= 45·58$

16 $52 \div 10 =$ **5·2** $89 \div 100 =$ **0·89** $7 \div 100 =$ **0·07**
4 $\div 10 = 0·4$ $88 \div 100 = 0·88$

17 Write each measure in decimal form.
$\frac{856}{1000}$ kg $\rightarrow$ **0·856 kg** $2\frac{25}{1000}$ ℓ $\rightarrow$ **2·025 ℓ** $6\frac{9}{1000}$ km $\rightarrow$ **6·009 km**

18 Write the numbers in order. Start with the smallest.

| 7·765 | 6·765 | 7·756 | 7·675 | 7·765 |

| 6·776 |

6·765   6·776   7·675   7·756   7·765

19 10% of 70 = **7**   40% of 300 = **120**   5% of 40 = **2**

20 Change 6 kg 452 g to grams. **6452 g**
Change 8080 g to kilograms and grams. **8 kg 80 g**

21 Find the distance in metres, between

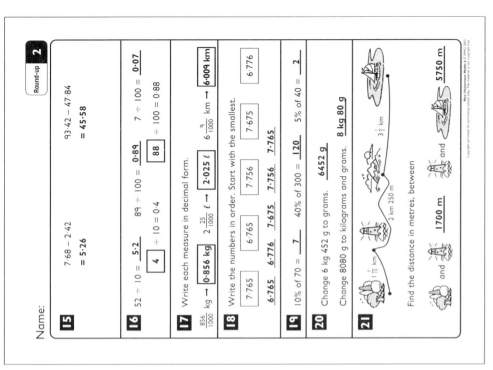

$1\frac{7}{10}$ km   2 km 250 m $3\frac{1}{2}$ km

____ and **1700 m**

____ and **5750 m**

Round-up 2

Name:

22 How many minutes are there between the **Start** and **Finish** times for each activity?

Activity	Start	Finish	
jogging	06.10	07.50	→ **100 min**
gardening	10.35	12.05	→ **90 min**
reading	20.15	22.00	→ **105 min**

23 Find the area of each right-angled triangle.

Area = **15 m²**   Area = **24 m²**

24

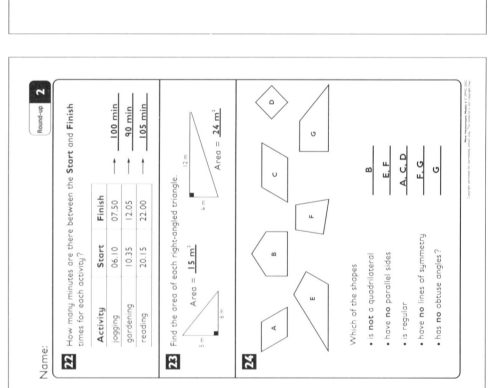

Which of the shapes
- is **not** a quadrilateral **B**
- have **no** parallel sides **E, F**
- is regular **A, C, D**
- have **no** lines of symmetry **F, G**
- has **no** obtuse angles? **G**

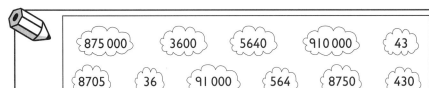

Home Activity
2

875 000   3600   5640   910 000   43

8705   36   91 000   564   8750   430

Use the clues to find each number in the panel above.

- 10 times this number is 56 400.   **5640**
- Divide this number by 100 and the result is 910.   **91 000**

- Multiply this number by 1000 and the product is 43 000.   **43**
- Divide this number by 10 and the result is 360.   **3600**

- 100 times this number is 870 500.   **8705**
- Divide this number by 1000 and the result is 875.   **875 000**

Complete each number chain.

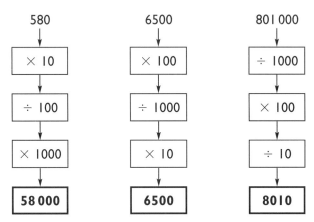

580 → × 10 → ÷ 100 → × 1000 → **58 000**

6500 → × 100 → ÷ 1000 → × 10 → **6500**

801 000 → ÷ 1000 → × 100 → ÷ 10 → **8010**

Home Activity
3

$26 + 18 + 34 + 22 =$ **100** $13 + 42 + 25 + 24 =$ **104**

$84 + 62 + 46 =$ **192** $97 + 39 + 43 =$ **179**

$67 + 43 + 23 + 64 =$ **197** $56 + 99 + 25 + 39 =$ **219**

$38 +$ **51** $+ 61 = 150$ $74 + 45 +$ **61** $= 180$

How much greater is the total of the numbers in the circle than the total of the numbers in the square?   **10**

62 66
64 61
65

79 76
75 78

Home Activity
4

$6400 + 2500 =$ **8900** $1200 + 8700 =$ **9900**

$3600 + 4800 =$ **8400** $2300 + 4900 =$ **7200**

$4600 +$ **4300** $= 8900$ $2600 +$ **3700** $= 6300$

$4200 + 9700 =$ **13 900** $2700 + 8500 =$ **11 200**

$5600 +$ **6700** $= 12300$ $7700 +$ **7400** $= 15100$

Home Activity 4

6400 + 2500 = **8900**   1200 + 8700 = **9900**

3600 + 4800 = **8400**   2300 + 4900 = **7200**

4600 + **4300** = 8900   2600 + **3700** = 6300

4200 + 9700 = **13 900**   2700 + 8500 = **11 200**

5600 + **6700** = 12300   7700 + **7400** = 15100

Home Activity 5

450 − **170** = 280   760 − **390** = 370   810 − **550** = 260

320 − 140 = 180   **630** − 360 = 270   **210** − 120 = 90

537 − 160 = **377**   748 − 270 = **478**   629 − 380 = **249**

Home Activity 6

13 200 − 8800 = **4400**   11 400 − 9700 = **1700**

3300 − **1500** = 1800   **5400** − 2900 = 2500

7900 − 5150 = **2750**   6800 − 3250 = **3550**

8500 − 2350 = **6150**   4600 − 2550 = **2050**

3700 − 1050 = **2650**   10 000 − 3450 = **6550**

Home Activity 8

 63p   48p   37p

Find the cost of

3 comics £1·89   6 cans of juice £2·88

7 packets of crisps £2·59   8 comics £5·04

4 cans of juice £1·92   9 packets of crisps £3·33

Home Activity 9

$101 \times 48 = \underline{\textbf{4848}}$ $27 \times 99 = \underline{\textbf{2673}}$ $24 \times 101 = \underline{\textbf{2424}}$

$66 \times 99 = \underline{\textbf{6534}}$ $99 \times 13 = \underline{\textbf{1287}}$ $101 \times 75 = \underline{\textbf{7575}}$

What is the cost of

19 Train sets   £1881

26 Castle kits?   £2626

Train Set   £99

Castle Kit   £101

Home Activity 11

Halve each number.
Match each **answer** to a letter to find the secret message.

624	W		288	E		1240	L		4800	L

606	D		8400	O		1760	N		964	E

6800	B		9200	R		562	A		718	I		1130	N

1076	B		398	O		7400	X

D	B	R	E	N	O	L
303	3400	4600	482	565	199	620
B	**L**	**P**	**W**	**I**	**S**	**X**
538	2400	830	312	359	432	3700
E	**R**	**U**	**A**	**M**	**O**	**N**
144	720	532	281	154	4200	880

Home Activity 12

Complete each number sequence and write its rule.

$\underline{3}$, $\underline{18}$, $\underline{33}$, **48**, **63**, $\underline{78}$ ⟶ **Add 15**

$\underline{112}$, $\underline{91}$, $\underline{70}$, **59**, **38**, $\underline{7}$ ⟶ **Subtract 21**

$\underline{9}$, **34**, **59**, $\underline{84}$, $\underline{109}$, $\underline{134}$ ⟶ **Add 25**

$\underline{99}$, $\underline{80}$, **61**, **42**, $\underline{23}$, $\underline{4}$ ⟶ **Subtract 19**

61, $\underline{50}$, $\underline{39}$, **28**, $\underline{17}$, $\underline{6}$ ⟶ **Subtract 11**

$\underline{^-18}$, $\underline{^-12}$, $\underline{^-6}$, **0**, **6**, $\underline{12}$ ⟶ **Add 6**

$\underline{64}$, $\underline{32}$, **16**, $\underline{8}$, $\underline{4}$, **2** ⟶ **Halve the number**

Home Activity 13

Complete.

$1 + 3 = \underline{\textbf{4}}$

$1 + 3 + 5 = \underline{\textbf{9}}$

$1 + 3 + 5 + \underline{\textbf{7}} = \underline{16}$

$1 + 3 + \underline{\textbf{5}} + \underline{\textbf{7}} + \underline{\textbf{9}} = \underline{25}$

$1 + 3 + \underline{\textbf{5}} + \underline{\textbf{7}} + \underline{\textbf{9}} + \underline{\textbf{11}} = \underline{36}$

$1 + 3 + \underline{\textbf{5}} + \underline{\textbf{7}} + \underline{\textbf{9}} + \underline{\textbf{11}} + \underline{\textbf{13}} = \underline{49}$

What do you notice?   **The answers are all square numbers.**

$0.37 + 0.41 =$ __**0·78**__ $\quad$ $0.52 + 0.26 =$ __**0·78**__

$0.5 + 0.82 =$ __**1·32**__ $\quad$ $0.45 + 0.8 =$ __**1·25**__

$1.34 + 2.15 =$ __**3·49**__ $\quad$ $3.26 + 1.31 =$ __**4·57**__

$4.63 + 2.07 =$ __**6·7**__ $\quad$ $5.61 + 1.26 =$ __**6·87**__

$6.13 +$ **0·87** $= 7$ $\quad$ **0·07** $+ 6.13 = 6.2$

$2 \times 8.7 =$ **17·4**	$4.5 \times 3 =$ **13·5**	$4 \times 8.6 =$ **34·4**
$9.1 \times 5 =$ **45·5**	$8 \times 3.2 =$ **25·6**	$4 \times 7.8 =$ **31·2**
$5.7 \times 3 =$ **17·1**	$7.8 \times 2 =$ **15·6**	$4.1 \times 9 =$ **36·9**
$6 \times 5.2 =$ **31·2**	$7 \times 8.6 =$ **60·2**	$2.4 \times 8 =$ **19·2**

$\frac{1}{10}$ of $700 =$ **70** $\quad$ $\frac{1}{7}$ of $63 =$ **9** $\quad$ one hundredth of $800 =$ **8**

$\frac{2}{5}$ of $35 =$ **14** $\quad$ $\frac{3}{8}$ of $48 =$ **18** $\quad$ seven ninths of $45 =$ **35**

$\frac{1}{3}$ of $93 =$ **31** $\quad$ $\frac{5}{6}$ of $66 =$ **55** $\quad$ five sevenths of $280 =$ **200**

$\frac{9}{10}$ of $400 =$ **360** $\quad$ $\frac{8}{9}$ of $360 =$ **320** $\quad$ four sixths of $420 =$ **280**

$\frac{3}{5}$ of $2 \text{ kg} =$ __**1·2 kg**__ $\quad$ $\frac{7}{10}$ of £3 $=$ __**£2·10**__ $\quad$ $\frac{7}{8}$ of $6 \text{ m} =$ __**5·25 m**__

Help
- To find $\frac{3}{4}$ of 28, work out **one** quarter ($28 \div 4 = 7$) then multiply by 3 ($7 \times 3 = 21$).
- To find a fraction of a quantity, first change kg to g, £ to p and m to cm, for example: $\frac{2}{5}$ of £4 → $\frac{2}{5}$ of 400p.

Write each decimal as a fraction.

1	6	•	2		$16\frac{2}{10}$
8	•	9	7		$8\frac{97}{100}$
2	•	6	5		$2\frac{65}{100}$
4	3	•	8		$43\frac{8}{10}$

	1	•	3	9	$1\frac{39}{100}$
3	7	•	4		$37\frac{4}{10}$
	4	•	0	2	$4\frac{2}{100}$
7	5	•	5		$75\frac{1}{2}$

Home Activity 20

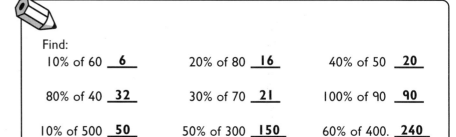

$5 \div 10 = \underline{\textbf{0·5}}$ $16 \div 10 = \underline{\textbf{1·6}}$ $87 \div 10 = \underline{\textbf{8·7}}$

$62 \div 100 = \underline{\textbf{0·62}}$ $31 \div 100 = \underline{\textbf{0·31}}$ $8 \div 100 = \underline{\textbf{0·08}}$

$\boxed{43} \div 10 = 4·3$ $\boxed{94} \div 100 = 0·94$ $\boxed{2} \div 100 = 0·02$

Use Home Sheet 27   **Home Activity 21**

Write the decimal fraction shown by each arrow.

0·39 **0·40** 0·41 0·42 0·43 0·44 0·45 0·46 0·47 0·48 0·49 **0·50** 0·51

0·394 $\dfrac{408}{1000}$ $\dfrac{431}{1000}$ 0·455   0·479 $\dfrac{503}{1000}$

7·79 **7·80** 7·81 7·82 7·83 7·84 7·85 7·86 7·87 7·88 7·89 **7·90** 7·91

$\dfrac{7797}{1000}$ 7·820 $\dfrac{7844}{1000}$ $\dfrac{7876}{1000}$ 7·892 7·902

Draw and label arrows to show:

0·394   7·892   0·455   7·902   0·479   7·820

Home Activity 22

Find:

10% of 60 __**6**__	20% of 80 __**16**__	40% of 50 __**20**__
80% of 40 __**32**__	30% of 70 __**21**__	100% of 90 __**90**__
10% of 500 __**50**__	50% of 300 __**150**__	60% of 400. __**240**__

Home Activity 23

Write each time as a 12-hour time. Use am or pm.

16.45 **4.45 pm**   07.10 **7.10 am**

23.25 **11.25 pm**   11.55 **11.55 am**

Write each time as a 24-hour time.

6.50 am __**06.50**__   8.20 pm __**20.20**__

1.05 pm __**13.05**__   12.30 am __**00.30**__

Write these times in order. Start with the **latest**.

12.05   11.50   11.45pm   21.30   12.10 am   09.35

__**11.45 pm**__, __**21.30**__, __**12.05**__, __**11.50**__, __**09.35**__, __**12.10 am**__